Ernesto Sanchez, Giovanni Squillero, and Alberto Tonda

Industrial Applications of Evolutionary Algorithms

Intelligent Systems Reference Library, Volume 34

Editors-in-Chief

Prof. Janusz Kacprzyk
Systems Research Institute
Polish Academy of Sciences
ul. Newelska 6
01-447 Warsaw
Poland
E-mail: kacprzyk@ibspan.waw.pl

Prof. Lakhmi C. Jain
University of South Australia
Adelaide
Mawson Lakes Campus
South Australia 5095
Australia
E-mail: Lakhmi.jain@unisa.edu.au

Further volumes of this series can be found on our homepage: springer.com

Vol. 10. Andreas Tolk and Lakhmi C. Jain
Intelligence-Based Systems Engineering, 2011
ISBN 978-3-642-17930-3

Vol. 11. Samuli Niiranen and Andre Ribeiro (Eds.)
Information Processing and Biological Systems, 2011
ISBN 978-3-642-19620-1

Vol. 12. Florin Gorunescu
Data Mining, 2011
ISBN 978-3-642-19720-8

Vol. 13. Witold Pedrycz and Shyi-Ming Chen (Eds.)
Granular Computing and Intelligent Systems, 2011
ISBN 978-3-642-19819-9

Vol. 14. George A. Anastassiou and Oktay Duman
Towards Intelligent Modeling: Statistical Approximation Theory, 2011
ISBN 978-3-642-19825-0

Vol. 15. Antonino Freno and Edmondo Trentin
Hybrid Random Fields, 2011
ISBN 978-3-642-20307-7

Vol. 16. Alexiei Dingli
Knowledge Annotation: Making Implicit Knowledge Explicit, 2011
ISBN 978-3-642-20322-0

Vol. 17. Crina Grosan and Ajith Abraham
Intelligent Systems, 2011
ISBN 978-3-642-21003-7

Vol. 18. Achim Zielesny
From Curve Fitting to Machine Learning, 2011
ISBN 978-3-642-21279-6

Vol. 19. George A. Anastassiou
Intelligent Systems: Approximation by Artificial Neural Networks, 2011
ISBN 978-3-642-21430-1

Vol. 20. Lech Polkowski
Approximate Reasoning by Parts, 2011
ISBN 978-3-642-22278-8

Vol. 21. Igor Chikalov
Average Time Complexity of Decision Trees, 2011
ISBN 978-3-642-22660-1

Vol. 22. Przemysław Różewski, Emma Kusztina, Ryszard Tadeusiewicz, and Oleg Zaikin
Intelligent Open Learning Systems, 2011
ISBN 978-3-642-22666-3

Vol. 23. Dawn E. Holmes and Lakhmi C. Jain (Eds.)
Data Mining: Foundations and Intelligent Paradigms, 2011
ISBN 978-3-642-23165-0

Vol. 24. Dawn E. Holmes and Lakhmi C. Jain (Eds.)
Data Mining: Foundations and Intelligent Paradigms, 2011
ISBN 978-3-642-23240-4

Vol. 25. Dawn E. Holmes and Lakhmi C. Jain (Eds.)
Data Mining: Foundations and Intelligent Paradigms, 2011
ISBN 978-3-642-23150-6

Vol. 26. Tauseef Gulrez and Aboul Ella Hassanien (Eds.)
Advances in Robotics and Virtual Reality, 2011
ISBN 978-3-642-23362-3

Vol. 27. Cristina Urdiales
Collaborative Assistive Robot for Mobility Enhancement (CARMEN), 2011
ISBN 978-3-642-24901-3

Vol. 28. Tatiana Valentine Guy, Miroslav Kárný and David H. Wolpert (Eds.)
Decision Making with Imperfect Decision Makers, 2012
ISBN 978-3-642-24646-3

Vol. 29. Roumen Kountchev and Kazumi Nakamatsu (Eds.)
Advances in Reasoning-Based Image Processing Intelligent Systems, 2012
ISBN 978-3-642-24692-0

Vol. 30. Marina V. Sokolova and Antonio Fernández-Caballero
Decision Making in Complex Systems, 2012
ISBN 978-3-642-25543-4

Vol. 31. Ludomir M. Laudański
Between Certainty and Uncertainty, 2012
ISBN 978-3-642-25696-7

Vol. 32. José J. Pazos Arias, Ana Fernández Vilas, and Rebeca P. Díaz Redondo
Recommender Systems for the Social Web, 2012
ISBN 978-3-642-25693-6

Vol. 33. Jie Lu, Lakhmi C. Jain, and Guangquan Zhang
Handbook on Decision Making, 2012
ISBN 978-3-642-25754-4

Vol. 34. Ernesto Sanchez, Giovanni Squillero, and Alberto Tonda
Industrial Applications of Evolutionary Algorithms, 2012
ISBN 978-3-642-27466-4

Ernesto Sanchez, Giovanni Squillero, and Alberto Tonda

Industrial Applications of Evolutionary Algorithms

Springer

Author

Prof. Ernesto Sanchez
Politecnico di Torino - DAUIN
Italy

Dr. Alberto Tonda
Politecnico di Torino - DAUIN
Italy

Prof. Giovanni Squillero
Politecnico di Torino - DAUIN
Italy

ISSN 1868-4394 e-ISSN 1868-4408
ISBN 978-3-642-27466-4 e-ISBN 978-3-642-27467-1
DOI 10.1007/978-3-642-27467-1
Springer Heidelberg New York Dordrecht London

Library of Congress Control Number: 2011945155

Printed on acid-free paper

Springer is part of Springer Science+Business Media (www.springer.com)

Foreword

My mother was a very simple plain person, with a hearth of gold. She was proud of my academic successes and research career; she often asked me "But what does your work consist of?". I am an expert in Computational Intelligence, i.e., neural networks, fuzzy logic and evolutionary computation, and although I am a full professor at the Faculty of Engineering of the University of Pisa, it was not easy to answer my mother's question. She used to think that her world ("the real world") and my world ("the artificial, intellectual world") were so far from each other that no means could ever exist to let these worlds interact. But one day, by chance, evolutionary computation made the miracle happen! Yes, that day I went to visit my mother and found her completely busy doing a lot of things. So I proposed her a sort of game. I said to her: "You have to perform a set of tasks (such as cleaning up the house, cooking the dinner, etc), each consisting of a series of more elementary operations, which can be performed in parallel or sequentially. You need an operation sequence plan that specifies the collection and the order of operations to carry out. Actually, not all the combinations of operations are feasible, e.g., you cannot start cooking the pasta before putting a saucepan on to cook. Further, some feasible and valid sequences can be better, e.g., less time-consuming or demanding fewer tool changes, than others. Of course your desire is to finish your work as good and as early as possible. How do you choose the operation plan?"

"I do not follow any rule, it is just habit", she answered.

"Then, consider all the single operations making the specific tasks to perform, and write down on a piece of paper a few randomly-generated sequences of these operations. Now rank the generated operation sequences based, e.g., on feasibility and time/fatigue effort requirements. If the best sequence satisfies your desire, follow that operation plan. Otherwise let an Evolutionary Algorithm (EA) generate a new list of operation sequences obtained by automatically combining the current sequences in an appropriate way. Check if the best sequence is good for you, otherwise repeat the process again, and so on. You can be sure that the EA will find a good solution sooner or later."

She looked at me astonished, without speaking. I went on saying "This is a simple way to mimic natural evolution, and EAs do exactly this in an automatic way."

What had happened that day? I and my mother had simply managed to make our worlds interact, a thing that appeared impossible up to that moment. It is like Columbus's egg.

Probably this is the right and only way to fill the gap between apparently different worlds like that of modern industrial applications and that of EAs. EAs can perform systematic random search in order to improve the likelihood of finding globally optimal solutions. On the other hand experience has shown that awareness of real-world industrial problems and knowledge of traditional computation techniques are not always enough to cope with the growing complexity of modern industrial processes and products. Then, why not to use the potentiality of EAs? Probably industrial experts are simply not aware of how EAs could be applied to solve their problems. In fact the key point to applying EAs to solve otherwise intractable problems is just representing and assessing the candidate solutions to a problem in an appropriate way.

It is just like Columbus's egg. Let the experts of EAs show industrial engineers and operators what EAs can do! The current book makes exactly this by presenting a collection of real significant industrial problems and their EA-based solutions. The considered case studies help the reader learn to employ EAs with a minimal investment in time and effort. This is what makes the current book useful and valuable for effective technology transfer into industrial organizations. Described applications include automatic software verification, test program generation for microprocessors, test generation for hardware and circuits, antenna array synthesis and optimization, drift correction of chemical sensors, and generation of test sets for on-line test of microprocessors.

Now that EAs represent a pretty mature field this is the right book for all postgraduates, research scientists and practitioners who want to tackle challenging industrial problems, of whatever complexity, with the most up-to-date, powerful and easy-to-use optimization technology.

Pisa, Italy, September 2011 *Beatrice Lazzerini*

Preface

The increasing complexity of products and processes leads directly to the growing intricacy of the problems and issues the industrial world is facing. More and more often, traditional computational techniques prove unable to cope with real world situations, either because the time needed to reach an optimal solution is not compatible with the frantic development processes of a company, or because the modeling of complex systems to the degree of precision needed is unfeasible. *Evolutionary Algorithms* (EA) comprehend a wide class of stochastic bio-inspired optimization techniques, firstly developed by J. H. Holland, L. J. Fogel, I. Rechenberg and H. Schwefel during the late 1960s and early 70s. Over the course of the last 35 years, EAs demonstrated their effectiveness in an extended variety of problems, ranging from airfoil design to credit card fraud detection. The industrial world, however, is still reluctant to introduce these powerful techniques into real procedures, mainly due to the sensation of insufficient controllability, scarce repeatability of the results, and the lack of experts with deep knowledge of both EAs and modern industrial needs. This book presents different case studies of EAs successfully applied to real world problems, hopefully showing the untapped potential of these techniques in various industrial fields.

Chapter 1 comprehends a description of typical complex industrial problems, a brief history of EAs, a comparison with traditional methods, and a discussion on the application of evolutionary techniques to real world problems.

Chapter 2 presents what is meant to be a surely incomplete, but extremely useful list of resources relevant for further elaboration and understanding of the multifaceted world of EAs.

The first section groups industrial problems related to the verification of hardware and software working prototypes.

The case study presented in chapter 3 deals with the software verification of a whole operative system and all applications running on a mobile phone prototype. The chapter focus specifically on the problems concerning the application of an EA to a "needle in a haystack" kind of problem; on how to make the EA perform; and on how EAs can complete human expertise in the software verification field. The activity is carried out in cooperation with Motorola Research Labs, Torino, Italy.

The verification of microprocessors is a growing field of study, mainly because design capability outperforms current verification techniques. Most studies on the correct behavior of a microprocessor are thus run on working prototypes, in the attempt to locate critical paths by making the device fail its computations. In chapter 4, an EA-based method to identify critical speed-paths in a multi-core microprocessor, exceeding the performance of state-of-the-art stress tests, is described.

The second section presents a collection of real-world case studies pertaining design and reliability.

The design of an antenna array is the topic of chapter 5. When devising such a complex system, often manual or automatic optimization methods do not yield satisfactory results, being either too labour-intensive or unsuitable for some specific class of problems. When an evolutionary algorithm is used to optimize parameters of the antenna array, the results show that these techniques are able to obtain better results than both manual and automatic approaches.

In chapter 6, an EA-based technique to lengthen the lifespan of *electronic noses*, complex olfactory sensor arrays, is presented. Sensor arrays are affected by the *drift* problem, a degenerative error in the measurements, hard to model and predict. The proposed solution is to dynamically adjust the sensor readings with a state-of-the art Evolutionary Strategy proves to be effective. The experience is performed with the collaboration of Sensor CNR-INFM Lab, Brescia, Italy.

Chapter 7 tackles the problem of automatically devising online test sets for microprocessor-based systems. While existing manufacturing test set can be used for this purpose, several additional constraints must be considered for an online application, including test length, duration, and intrusiveness. The proposed methodology, here applied to an Intel 8051 microcontroller, exploits an EA to create online test sets starting from tests devised by the manufacturer.

The third section introduces results obtained through the application of EAs to test generation problems for hardware and circuits.

Chapter 8 concerns the study of path delay faults in electronic devices, misbehaviors where a device produces a correct result without conforming to time specifications. Devising test to uncover the presence of these faults is challenging, exspecially when only a high-level description of the device is provided. To tackle this problem, where the ideal result is a set of equally feasible solutions, a Multi-Objective Evolutionary Algorithm (MOEA) is employed.

In chapter 9, EAs are applied to the field of Software-Based Self Testing (SBST), an established test technique for various hardware architectures. SBST's principle is to apply a suitable series of stimuli to the device under test, comparing the produced output to the expected one. Finding a minimal set of stimuli to thoroughly excite a device is not a trivial problem: EAs prove successful once again, showing that the proposed methodology is effective on a wide range of hardware peripherals.

Chapter 10 deals again with stimuli generation for SBST, this time tackling a much more complex system, such as a microprocessor pipeline. Using a high-level representation of the target device, and a dynamically built Finite State Machine (FSM), fault coverage of the candidate stimuli are evaluated without resorting to time-expensive simulations on low-level models. Experimental results show that the evolved test obtains a nearly complete fault coverage against the targeted fault model.

Acknowledgments

The authors would like to express their gratitude towards their families and colleagues for their invaluable support, useful ideas and intriguing discussion. A particular thank to A. Aimo, P. Bernardi, A. Cerato, K. Christou, S. Di Carlo, S. Drappero, M. Falasconi, G. Fisanotti, M. Grosso, S. Loiacono, M. K. Michael, L. Manetta, A. Moscatello, L. Motta, L. Ollino, D. Ravotto, T. Rosato, W. Ruzzarin, M. Schillaci, A. Scionti and M. Sonza Reorda; without their help, this book would have not been possible.

Contents

Chapter 1
Introduction

This first chapter provides the reader with a survey of current industrial problems, hinting at their complexity and variety. It is shown how traditional computational techniques often fail to deliver the expected results in modern real-world applications, while computational intelligence show an increasing amount of interesting results. Some background and a brief history of Evolutionary Algorithms (EAs) are then provided, introducing these interesting stochastic optimization techniques.

1.1 Industrial Problems

In the modern industrial world, the complexity of problems faced by companies is growing accordingly with the complexity of products they sell. This holds particularly true for the IT field: considering hardware components, the number of connections per silicon wafer is doubling each year, following closely the famous "Moore's Law"; and with a more and more inexpensive and powerful hardware at disposal, the number of applications manageable by operative systems, even on mobile devices, is increasing at almost the same rate.

Such a growth in complexity directly leads to difficulties in every step of product development, starting from the design step. Even determining the correct combination of parameters to obtain the desired behavior for a device is not a straightforward process, because each choice could have intricate and sometimes not foreseeable repercussions. More and more, industry must resort to heuristic and meta-heuristic techniques to find the best alternative between different possibilities. Evolutionary algorithms proved successful in solving several design-related issues, from antennas optimization [101] to fine-tuning of product details to maximize its recycling possibilities [163].

Verification and testing, for which considerable amounts of time and money are invested during the development of a new product, are also heavily influenced by the increasing complexity of devices: microprocessors' designers, for example, candidly acknowledge that "very few chips ever designed function or meet their

E. Sanchez et al.: Industrial Applications of Evolutionary Algorithms, ISRL 34, pp. 1–10.
springerlink.com

performance goal the first time" [106]. In practice, production capacity outperforms testing capacity by several orders of magnitude, and activities once performed on models are now applied to physical prototypes, despite the enormous costs involved in prototyping.

When the single parts of a device become so intricate, the interaction between them could also lead to extreme difficulties in making predictions on the behavior and the lifespan of the device itself. Time to market is a pressing issue for the industry, and the possibilities of studying throughly a system are often limited. Issues in the functionalities of a product can arise in an unexpected way, at unexpected moments: thus, the necessity arises for ways to solve unexpected problems when they appear.

Even when reliable models of devices are available, developing efficient verification sets is not trivial. Often the methodologies applied must adhere to strict constraints of time and memory occupation, since their results could be used under different conditions. In other contexts, some crucial parts of a device are so embedded that even observing the tests' result becomes a non-trivial activity. An example are peripherals in Systems-on-Chip, devices that integrate all components of a computer or other electronic system into a single integrated circuit.

When the product is a software application, although development steps are significantly different, the call for complexity is still in place, transposed to algorithmic level. Image analysis, for example, is one of the fields where traditional techniques are lagging behind: computational intelligence, on the other hand, showed promising results in complicate tasks such as fractal approximation [37], pattern recognition [24] and tracking of moving objects in videos [117].

Classification and data mining also suffer from the increment of available data: in a heavily connected world, statistics obtained from social networks and other websites can be of great interest for companies which desire to advertise their products to a specific audience. Identifying meaningful patterns in such a huge amount of information, however, is a hard problem, and even here classical approaches are showing their limits and the first applications of computational intelligence are appearing in fields such as credit card fraud detection [19].

Several other interesting industrial applications of computational intelligence can be found among the case studies reported in [163], ranging from design of optical fibers to optimization of store performances, from planning of railway track intervention to applications in chemical industries.

1.2 A Brief History of Evolutionary Algorithms

The *Theory of evolution* postulates that all living organisms have their origin in other preexisting beings: differences between current lifeforms have their origin in modifications inherited through successive generations. *Evolutionary computation* is the branch of Computer Science that focuses on algorithms inspired by the natural world and the theory of evolution: while this definition may seem vague, the field of study has boundaries that are not, and cannot be, defined clearly. Evolutionary computation is included in the broader group of *bio-inspired heuristics*, which are in turn a sub-section of *computational intelligence*. In the following, the distinction between *natural evolution* and *artificial evolution* will be stressed out for clarity whenever necessary.

This section is meant to be a brief presentation of the basics of evolutionary computation and its terminology: a thorough description of the topic is out of the scope of this book, and most concepts are detailed only to the extent they are required in the following. Readers interested in a comprehensive coverage of the field will find several fascinating books on the topic, for example [56]. For a survey of the vast and alluring world of biology studies, [48] and [66] can be two interesting starting points.

1.2.1 Natural and Artificial Evolution

Scientists show a remarkable consensus on the theory of natural evolution, which is considered a cornerstone of modern biology. The current theory is the sum of several concepts: *evolution* and *natural selection* were introduced almost concurrently and independently by Charles Robert Darwin and Alfred Russel Wallace in 19th century; *selectionism* is an idea of Charles Weismann [160]; *genetics* have been analyzed first by Gregor Mendel [159]. This coherent corpus, often referred to as *Neo-Darwinism*, is able to explain the variety and characteristics of life on Earth starting from a limited number of relatively simple and plausible ideas: *reproduction*, *variation*, *competition*, and *selection*. In this context, reproduction is the process of generating an offspring from parents where the progeny inherit traits of their predecessors. Variation is the unexpected alteration of a trait. Competition and selection are the inevitable results of the strive for survival caused by an environment with limited resources.

By these concepts, evolution appears to be a set of random forces shaped by deterministic pressures: or, in other words, a sequence of steps, some mostly deterministic and some mostly random [105]. It is interesting to notice how similar ideas have been applied to describe phenomena not pertaining to biology, for example alternatives conceived during learning [26], ideas striving to survive in our culture [48], or even possible universes [167] [143].

Eminent biologists, such as Richard Dawkins and Stephen Jay Gould in recent times, repeatedly warned the non-specialist community against mistaking evolution for a process of improvement or optimization, going from raw to perfected features. Nevertheless, assuming for an instant that evolution *is* in fact a force pushing towards an objective, its results are astonishing: over the course of billion years, it turned unorganized groups of cells into startlingly complex structures such as wings and eyes, without the need of any a-priori design. Following this idea, the neo-Darwinistic paradigm itself can be seen as an effective optimization tool, producing great results from scratch, advancing without a plan, exploiting a mix of random and deterministic techniques.

Setting aside biologists' warnings, evolutionary computation makes use of these powerful ideas to search for optimal solutions in various sets of problems. All these problems often have one common feature: the best way to reach the optimum is not known, at least not in detail. By exploiting neo-Darwinian principles, sets of candidate solutions are cultivated in artificial environments, modified in discrete steps, and selected by an environment defined by the characteristics of the problem itself. Good solutions at a given step inherit positive traits from their ancestors, and optimal results eventually arise from the artificial primordial soup. Unlike evolution, this process has a precise goal; also, these simulated evolutions are often simplified to the extent that they become unrealistic: nevertheless, scientific literature routinely reports success stories of evolutionary computation applied to a vast number of fields.

A small set of terms specific to evolutionary computation is now introduced, since they will be consistently used in the following chapters. Most of the terminology follows closely that of biology. A single candidate solution to a considered problem is called *individual*; a *population* is a group of individuals; and each step of the simulated evolution is termed a *generation*. The *fitness function* measures the effectiveness of an individual in solving the problem. Individuals with high fitness values are more likely to propagate their characteristics to the next generation. The word *genome* always denotes all the genetic information of the individual, even if different approaches use different techniques to store and manage this data. The smallest fragment of the genome that can be modified during the evolution is called *gene*: a gene can also be seen as the functional unit of inherited characteristics. The specific position where a gene is placed in the genome is known as *locus* (plural *loci*). The alternative genes that may appear in a given locus are called *alleles*.

While in biology there is a significant difference between *genotype*, all the genetic material of an organism, and *phenotype*, the observable characteristics that emerge from the interaction between the genotype and its environment, in evolutionary computation this distinction is often disregarded. Genotype and phenotype often coincide, even if sometimes the numerical value representing the fitness of an individual is assimilated to its phenotype.

When the offspring of individuals in evolutionary computation must be produced, often the algorithms exploit the paradigms of sexual and asexual reproduction in nature. Sexual reproduction is usually referred to as *recombination* or *crossover*: the resulting individual will inherit different characteristics from two or more parent individuals. Asexual reproduction is named *replicaton* or *mutation*: a copy of a parent individual is created and slightly altered. Some implementations consistently combine the two approaches, using mutation only after sexual recombination. Very few evolutionary algorithms assign distinct reproductive roles to individuals, so gender is almost never taken into account. Other implementations do not store a collection of individuals, but only a set of statistical parameters that describe the current population: in that case reproduction is performed by altering the parameters. All algorithmic techniques used to model natural reproduction can be called *evolutionary operators* or *genetic operators*, since they influence the genotype of individuals.

It is clear how variability in the evolutionary process is introduced by mutation and recombination; parent selection also uses a stochastic approach, even if often weighted by the fitness values of individuals in the population, e.g. fittest individuals have a greater probability of being selected. In population-based evolutionary algorithms, the number of individuals in the system varies regularly at each generation: first, offspring is generated, adding new individuals to the population; then, the less fit individuals are discarded. This last step is deterministic: it models the struggle for survival in a hostile environment and it is often referred to as *survivor selection*, *selection* or *slaughtering*.

Evolutionary algorithms may be classified as local search algorithms, since they explore a portion of the search space which is dependent on their actual state, with the offspring loosely defining the concept of neighborhood. Also, since they make use of a *trial and error* paradigm, and that they are not usually able to mathematically guarantee to find an optimal solution in a finite amount of time, evolutionary algorithms can be put into the group of heuristic algorithms: over the years, however, experts of the field have demonstrated the presence of several useful mathematical properties in their processes.

It is noticeable how the definition of evolutionary computation has no clear boundaries, and this branch of computational intelligence also lacks a single recognizable origin. In 1950, the great computer scientist Alan Turing was probably the first to point out the similarities between the processes of learning and evolution [154]. Near the end of the same decade, inspiring ideas in that direction began to appear [61] [111] [18], even if their diffusion among the broader scientific community was blocked by the lack of computational power available at the time. While some scholars point at this time frame as the origin of evolutionary computation, most of them agree that its birth is to be placed in the 1960s, with the appearance of three independent research lines: *genetic algorithms*, *evolutionary computation*, and *evolution strategies*. While an unanimous consensus on the matter is hard to reach, the fundamental importance of these contributions is unquestionable. A fourth paradigm, that appeared in the 1980s, must be also considered for both its novelty and its closeness to the aforementioned ideas.

1.2.2 Genetic Algorithms

Genetic algorithms (GA) are probably the most popular technique in evolutionary computation: they are so renowned that in non-specialized literature the term is sometimes used to denote any kind of evolutionary algorithm. John Holland attested the importance of this paradigm in his 1975 book [75], but the methodology was used and described in previous years by a great number of researchers, including many of Holland's students [59] [60] [20]. In the beginning, genetic algorithms have been used as a step in the process of *classifier systems*, a technique also devised by

Holland, and they have been exploited more to study the mechanisms of evolution, than to solve actual problems. In the first experiments a set of simple test benches, e.g. trying to set a number of bits to a specific value, were used to analyze different strategies and schemes.

In a genetic algorithm, an individual (i.e., the evolving entity), is represented as a sequence of bits. This is probably the only feature that was common to all the early implementations, while other choices may vary: the offspring produced at each step usually outnumbers the original population, various crossover and combination operators have been exploited by different researchers, and parents are often selected throughout a fitness-based probability distribution. During the selection of parents, highly fit individuals are favored by a bigger or smaller factor, depending on the *selective pressure* adopted in the algorithm. After the evaluation of new individuals, the population is shrank back to its original size. Several techniques to perform survivor selection have been used, but interestingly all methods to determine the survival of individuals are deterministic. Sometimes, all parents in the population are discarded, regardless of their fitness: if that cases, the approach is called *generational*. Conversely, if all individuals compete for survival independently from their age, the approach is named *steady-state*. All mechanisms that preserve the best individuals through subsequent generations fall under the scope of *elitism*.

1.2.3 Evolutionary Programming

Lawrence J. Fogel, in a series of works published at the beginning of 1960s [57] [58], proposed an algorithm that he called *evolutionary programming* (EP). The focus of Fogel's work was the evolution of predictive capabilities, since he was arguing that intelligent behavior requires the ability to forecast modifications in the environment: he used finite state machines (also called automata) as evolving entities, trying to evolve individuals able to anticipate the next symbol in an input sequence provided to them, thus showing a predictive capability. In later years, the same technique was successfully used to solve several combinatorial problems.

The original algorithm proposed by Fogel considered a set of P finite state machines: each individual in the set was tested against a sequence of symbols in input, i.e., its environment. The predictive capability was mapped to a single numerical value called fitness through different payoff functions that considered a penalty for too complex machines. Individuals were then ranked according to their fitness values, and subsequently P new automata were added to the population. Offspring generation was accomplished by mutation, whose type and extent were regulated by given probability distributions, so each new individual was obtained by modifying one existing automaton. In the first version of the algorithm, each selected parent created exactly one offspring, but the same automaton could be selected multiple times as the parent of different new individuals. Finally, half of the population was preserved and half discarded, so that its size returned to P. Survivors were chosen at random, with a probability related to their fitness value. The selective pressure in evolutionary programming is thus represented by the likeliness of a highly fit individual to be preserved in the next generation.

The steps described above were repeated until a specified number of generation had elapsed: at that moment, the best individual in the population was used to predict the actual next symbol, which was then added to the environment while the process restarted from the last population.

1.2.4 Evolution Strategies

Evolution strategies (ES) were proposed by Ingo Rechenberg and Hans-Paul Schwefel in the early 1960s [139] [128]. Originally developed as an optimization tool to solve practical problems, evolution strategies describe each individuals as a set of parameters, usually encoded as integer or real numbers. The mutation operator simultaneously modifies all parameters of a selected individual, with a high probability of inserting tiny alterations and a smaller probability of major modifications. On the other hand, several techniques are possible for the recombination operator: for example, copying a subset of parameters from each parent or computing an average of all the numbers. Interestingly, the very first experiment with evolution strategies featured a population of one individual and all random operations performed with a roll of six-sided dice.

A unique formalism was developed to describe the characteristics of evolution strategies. The Greek letter *mu* (μ) commonly denotes the size of the population, while for the size of the offspring generated at each generation the Greek letter *lambda* (λ) is used. A $(\mu + \lambda)$-ES is an evolution strategy where the offspring is added to the current population before survivor selection for the next generation. In this case, a particularly good solution could survive throughout several generations, as it could happen in a steady-state genetic algorithm or in evolutionary programming. The label (μ, λ)-ES, on the other hand, denotes an evolution strategy where the offspring completely replaces the current population before survivor selection. The latter approach has several similarities with a generational genetic algorithm or evolutionary programming, as the optimal solution may be discarded during the run. In a commonly used notation, the two approaches are called *plus*$(+)$ and *comma*$(,)$ respectively. These two terms spread in the evolutionary computation community, and in recent year they have been used in the description of various evolutionary algorithm, not necessarily related to evolution strategies. When comma selection is used, $\mu < \lambda$ must hold. In almost all the implementations of evolution strategies, however, the size of the offspring is much larger than the size of the population at each step.

When the recombination operator is implemented, the number of parents required is denoted with the Greek letter *rho* (ρ), and the algorithm with $(\mu/\rho \overset{+}{,} \lambda)$-ES. The number of parents is always smaller than the size of the population, i.e., $\rho < \mu$. $(\mu \overset{+}{,} 1)$-ES are sometimes referred to as *steady-state evolution strategies*.

An interesting approach, almost unique to evolution strategies, is *nesting*: instead of performing offspring generation with conventional operators, an evolution sub-strategy is started and its result is used as offspring for the main strategy. The inner strategy is in fact acting as a tool for local optimizations, and commonly it adopts parameters unrelated to those of the outer strategy. In different applications, this technique has been named *nested evolution strategies*, *hierarchical evolution strategies* and *meta evolution strategies*. An algorithm that exploits a sub-strategy running for γ generations is referred to as $(\mu/\rho \overset{+}{,} (\mu/\rho \overset{+}{,} \lambda)^{\gamma})$-ES, with γ is often called *isolation time*. A deeper nesting may be theoretically possible, but usually only one level of recursion is implemented. Such a technique is almost never used in evolutionary programming or genetic algorithms, but in rare cases has been successfully exploited in peculiar approaches [148].

Evolution strategies' offspring generation is mainly based on mutations: thus, different solution to determine the optimal amplitude of the perturbations were extensively explored during years of research. In real-valued search spaces, mutation is usually described as a random perturbation that follows a normal probability distribution centered on zero. In this way, small alterations are more probable than larger ones, while the variance may be used to tweak the average magnitude. This variance be evolved concurrently with individuals' parameters, and a dedicated variance may even be assigned to each parameter, since sometimes the same problem needs different amplitudes in different loci. In several implementations, this *variance vector* is modified using a fixed scheme, while the *object parameter vector*, i.e. the values that should be optimized, is modified using the variance vector: both vectors are then evolved concurrently as different parts of the same individual. This idea has been extended to take into account the correlation between optimal magnitudes of mutation, and modern evolution strategies often exploit a *covariance matrix*.

The capability to adapt to different problems is common to all evolutionary algorithms: thus, they can be sensibly called *adaptive*. When an evolutionary algorithm is able to adapt the mechanism of its adaptation, i.e., its internal parameters, is labeled as *self-adaptive*. Self-adapted parameters are sometimes called *endogenous*, from the term that describes hormones synthesized within an organism. Self-adaptation mechanisms have been exhaustively explored both in evolution strategies and evolutionary programming, and sometimes they appeared also in genetic algorithms.

From the 2000s there has been a growing interest in the use of evolution strategies as numerical optimization tools for continuous problems. Several versions of the most popular evolution strategies are freely available, with implementations ranging from general-purpose programming languages to commercial mathematical toolboxes, such as MatLab. Evolutionary programming also enjoyed a relatively widespread adoption as a numerical optimization tool, and the practical implementations of the two have mostly converged, even if the respective scientific communities remain deeply distinct.

This hybridization is not unique to evolution strategies and evolutionary programming. Ideas developed for one paradigm, if not directly applicable in other evolutionary algorithms, are at least a source of inspiration of the whole community. The various original approaches may be too different to interbreed, but many key ideas are now shared, and a great number of intermediate algorithms, not easily classifiable, have been described over the years. The scope of genetic algorithm also broadened, and researchers applied them to problems with highly structured solutions, e.g. the traveling salesman problem, whose solution is a permutation of nodes in a graph. The term genetic algorithm, however, remained strongly linked to the idea of bit strings of fixed length.

1.2.5 Genetic Programming

Genetic programming (GP) is the last paradigm appeared in the field of evolutionary computation, in order of time. John Koza, who applied for a patent in 1989, described this approach and made it popular in the community. The goal of the methodology is to automatically create computer programs, applying neo-Darwinistic concepts as optimization techniques. The first version of genetic programming was developed in *Lisp*, an interpreted computer language that dates back to the end of the 1950s. One of the characteristics of Lips it the capability to handle fragments of code as data, making it possible for a program to build up subroutines before evaluating them. Except variables and constants, everything in Lisp is treated as a prefix expression: since the first individuals in genetic programming were blocks of Lisp code, they were also prefix expressions. While the flexibility of Lisp has its advantages, the language is plagued by a severe inefficiency: during the development of genetic programming, researchers moved to alternative solutions, mostly featuring compiled languages. Since the origin of this paradigm, actually, the need for computational power and the effort to acquire efficiency have been important factors in the advancement of research. In these later implementations, the difference between a program and an expression became more evident than in Lisp: genetic programming algorithms in literature tackle mainly expressions.

Individuals in genetic programming are almost always internally represented as trees, despite the differences in the computer languages adopted. In the simplest forms, leaves (terminal nodes) encode numbers, while internal nodes describe operations. Variables, functions and programming structures appear in more complex variations. Offspring may be generated via recombination or mutation: the former is modeled as a swap of sub-trees between parents; the latter appears only in recent implementations, and usually involves a random modification of the tree, the promotion of a sub-tree to a new individual or the collapse of a sub-tree to a single terminal node. The first versions of genetic programming featured huge populations and made an extensive use of recombination, with a scarce presence of mutations: the substitution of a sub-tree may introduce a significant amount of novelty, since it is a potentially disruptive operation; while populations of significant size ensure that all possible symbols are available in the gene pool.

In recent years, many researchers have been attracted by the genetic programming paradigm. Various topics have been tackled, including representation of individuals, behavior of selection in numerous populations, techniques to avoid excessive growth of trees and different types of initializations. Results of these research lines has been used as test benches for practical techniques, or for foundation of theoretical studies. The genetic programming paradigm stimulated and brought new ideas to the whole evolutionary computation community.

Chapter 2
Resources

2.1 Books

T. Back, D. Fogel, Z. Michalewicz, *Handbook of Evolutionary Computation*, Oxford University Press, 1997

M. Bushnell, V. Agrawal, *Essentials of Electronic Testing for Digital, Memory, and Mixed-Signal VLSI Circuits*, Springer, 2000

D. Dumitrescu, B. Lazzerini, L.C. Jain, A. Dumitrescu, *Evolutionary Computation*, CRC Press, 2000

D. Fogel, *Evolutionary Computation: Toward a New Philosophy of Machine Intelligence*, Wiley, 2005

D. Gizopoulos (Ed.), *Advances in Electronic Testing*, Springer, 2006

D. Gizopoulos, A. Paschalis, A., T. Zorian, *Embedded Processor-Based Self-Test*, Springer, 2004

E. Sanchez, M. Schillaci, G. Squillero, *Evolutionary Optimization: the µGP toolkit*, Springer, 2011

J. D. Kraus, *Antennas*, McGraw-Hill Companies, 1988

J. W. Gardner, P. N. Bartlett, *Electronic Noses: Principles and Applications*, Oxford University Press, 1999

J. O. Smith III, *Introduction to Digital Filters: with Audio Applications*, W3K Publishing, 2007

E. Sanchez et al.: Industrial Applications of Evolutionary Algorithms, ISRL 34, pp. 11–13.
springerlink.com

2.2 Journals

Genetic Programming and Evolvable Machines
http://www.springer.com/computer/ai/journal/10710

IEEE Design and Test of Computers
http://www.computer.org/portal/web/dt

IEEE Transactions on Computers
http://www.computer.org/portal/web/tc

IEEE Transactions on Evolutionary Computation
http://www.ieee-cis.org/pubs/tec/

IEEE Transactions on Very Large Scale Integration Systems
http://www.ieee.org

Journal of Electronic Testing
http://www.springer.com/engineering/circuits+%26+systems/journal/10836

2.3 International Conferences and Workshops

CEC: Congress on Evolutionary Computation
<website changes every year>

DAC: Design Automation Conference
http://www.dac.com/

DATE: Design Automation & Test in Europe
http://www.date-conference.com/

EvoSTAR: The main European events on Evolutionary Computation
http://www.evostar.org/

GECCO: Genetic and Evolutionary Computation Conference
http://www.sigevo.org/gecco-XXXX/ (changes every year)

ITC: International Test Conference
http://www.itctestweek.org/about

2.4 Software

Eureqa
`http://creativemachines.cornell.edu/eureqa_download`

MicroGP (μGP)
`http://ugp3.sourceforge.net/`

Starting Points on the Web

`http://www.genetic-programming.com/`

`http://www.kesinternational.net/`

2.5 Suggested Readings on Natural Evolution and Biology

C. Darwin, *On the Origin of Species by Means of Natural Selection, or the Preservation of Favoured Races in the Struggle for Life*, John Murray, 1859

R. Dawkins, *The Selfish Gene*, Oxford University Press, 1976

R. Dawkins, *The Extended Phenotype*, Oxford University Press, 1982

S. Gould, *Ever Since Darwin*, W. W. Norton & Company Incorporated, 1977

S. Gould, *Punctuated Equilibrium*, Belknap Press of Harvard University Press, 2007

Part I
Prototype-Based Validation Problems

Chapter 3
Automatic Software Verification

The complexity of cell phones is continually increasing, with regards to both hardware and software parts. As many complex devices, their components are usually designed and verified separately by specialized teams of engineers and programmers. However, even if each isolated part is working flawlessly, it often happens that bugs in one software application arise due to the interaction with other modules. Those software misbehaviors become particularly critical when they affect the residual battery life, causing power dissipation. An automatic approach to detect power-affecting software defects is proposed. The approach is intended to be part of a qualifying verification plan and complete human expertise. Motorola, always at the forefront of researching innovations in the product development chain, experimented the approach on a mobile phone prototype during a partnership with Politecnico di Torino. Software errors unrevealed by all human-designed tests have been detected by the proposed framework, two out of three critical from the power consumption point of view, thus enabling Motorola to further improve its verification plans. Details of the tests and experimental results are reported.

3.1 Introduction

Verifying all the software running on a given apparatus is a complex problem, especially when the system under test is a mobile device, in which a software misbehavior can affect residual battery life. Traditional software verification techniques are often unable to work on a great number of applications at the same time, and since some software modules could be developed by third parties, verification engineers could not always have access to all data needed for the verification process. Evolutionary computation techniques proved able to tackle difficult problems with relevant degrees of success [43], even if some data of the problem is not completely known. Specialized literature routinely reports techniques that deliver high-return human-competitive machine intelligence simply starting from a high-level statement of what needs to be done and subsequently solving the problem without further need

E. Sanchez et al.: Industrial Applications of Evolutionary Algorithms, ISRL 34, pp. 17–30.
springerlink.com

of human intervention [86]. In the industrial world, however, the majority of existing processes employ no machine intelligence techniques, even if such approaches have been reported able to provide reliable results when facing complex problems.

The resistance in incorporating evolutionary computation in industrial processes may arise from the lack of experts with deep knowledge in both the machine intelligence and the industrial field. Automatic methodologies are perceived as scarcely controllable, and computational-intelligence techniques are regarded as "black magic", able to deliver impressive results sometimes, but definitely not reliable. In recent years, however, the interest of the industrial world towards automatic techniques has been steadily growing and some computational intelligence techniques have been successfully applied to some niche cases (e. g. credit card fraud detection performed by neural networks [19]).

An automatic approach based on an Evolutionary Algorithm (EA) is proposed, to add content to a human-designed verification plan for a mobile phone software system. The approach makes use of the EA to effectively [85] generate stimuli for a physical prototype of a cell phone, running simulations whose results are fed back to the EA and used to generate new stimuli. Data obtained from the simulations include physical measures and logs of all running applications. To explore effectively the solutions space, measures extracted from the prototype are integrated with data obtained from a model of the phone dynamically derived from simulation results.

Three different software misbehaviors, previously unrevealed by human-designed tests, are detected by the proposed approach. Incorporating this procedure in an existing set of tests allows Motorola [109] to further improve the effectiveness of qualifying verification plans. Preliminary results have been presented in [63] and [64].

3.2 Background

3.2.1 Mobile Phones

Since 1997, the mobile devices market has been steadily growing. Market researches projected that shipments of cell phones exceeded 1 billion units in 2009, so that mobile phones could become the most common consumer electronic device on the planet. Esteems from Gartner, shown in Fig. 3.1, predicted that there will be 2.6 billion mobile phones in use by the end of 2009 [161].

A great share of mobile devices sold nowadays is represented by the so-called smartphones, able to offer PC-like functionalities at the expense of an ever-growing complexity at both hardware and software level. Devices support more and more functions, running a great number of different applications: hardware miniaturization improves constantly, and thus battery life and power consumption related issues become more and more critical [36]. Thus, prediction of battery life [55] and

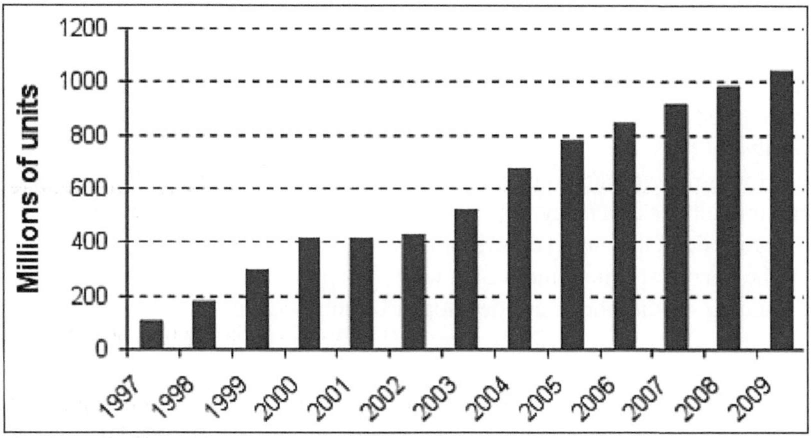

Fig. 3.1 Projection of cell phone sales by Gartner

improvement of energy supplies for mobile devices [90] [129] are research topics of great interest with significant contributions in literature.

Since the introduction of smartphones, the increasing number of applications run by mobile systems led to a great number of possible misbehaviors caused by software bugs. The most displeasing errors for the user are obviously those related to battery life, and in particular incorrect behaviors happening during the state where the cell phone consumes a minimal quantity of energy, called deep sleep. A mobile device enters deep sleep mode when it is left idle for a given amount of time or when a certain signal is given by the user (e. g. when the cap of a mobile phone is closed). Errors that arise in deep sleep can completely exhaust the battery of a cell phone while the user is oblivious to what is happening: a customer could find out that her mobile phone is discharged even if it was fully charged a few hours before.

3.2.2 Verification Techniques

Verification is the process that aims at guaranteeing the correctness of the design. Verification techniques exploit different paradigms, but, roughly speaking, it is possible to state that almost all can be classified either as formal or simulation-based. The former exploits mathematical methodologies to prove the correctness of the design with respect to a formal specification or property, while the latter is based on a simulation that aims at uncovering incorrect behaviors. Exploiting formal methods allows to verify the module with all possible inputs passing through all possible states. Therefore, these techniques in theory guarantee the highest levels of confidence in the correctness of the results, but when a formal method fails to prove a property, nothing can be determined about it, not even with a low amount of confidence. The human and computational effort required to apply formal verification techniques, severely limit their applicability. Such methods, as a result, are applied in the industrial field only when facing few software or hardware modules, when validation task can be significantly constrained by boundary conditions or when oversimplified models are employed, thus significantly impairing the confidence of the results [50]. Systems composed of a great number of modules usually cannot be tackled by formal verification, due to the growth of complexity of these techniques. To maximize their efficiency, formal verification techniques are usually applied to the source code of the model description. However, in the mobile phone prototyping arena, the very first time a mobile phone prototype is implemented, some applications running on the phone are developed by third parties and their original code is often non accessible [3]. Therefore, it is not always feasible to exploit formal verification techniques during the verification plan of a mobile phone.

Simulation-based techniques rely on the generation of a set of stimuli able to thoroughly excite the device under verification: the stimuli set is simulated exploiting the considered module. Subsequently, all data obtained from the simulation is gathered and analyzed, aiming to unearth misbehaviors by comparison with the expected results. A simulation-based approach may be able to demonstrate the presence of a bug even in frameworks with a great number of applications or hardware modules running simultaneously, but will never be able to prove its absence. Indeed, verification engineers may assume that no bugs exist depending on the level of confidence related to the quality of the simulated test set. Stimuli sets can be applied to either a physical prototype or a simulable model of the device. Both approaches have advantages and disadvantages: while models often describe only some aspects of the system, they may allow verification engineers to control all details of the simulation and gather a large amount of information. On the other hand, a physical prototype may be more difficult to control, but results of the physical emulation are completely unbiased, and the computational time required to apply the stimuli set is lower compared to model simulation. Either using a model or a prototype, the generation of a qualifying set of stimuli is the key problem with simulation-based techniques.

As mentioned by Piziali in [122], the real success of a simulation-based verification process relies on the adequacy of the initial verification route-map, called functional verification plan. A verification plan must define important test cases targeting specific functions of the design, and it must also describe a specific set of stimuli to apply to the design model. The verification plan can also allocate specific tasks to specialized engineers.

One of the most important tasks of the verification plan is the generation of the stimuli that thoroughly exercise the device, obeying the directives defined in the route-map.

According to the defined plan, different methodologies may be used to properly generate verification stimuli sets, for example deterministic, pseudo-random, or constrained-random. The generation of stimuli can be driven by past experience of the verification engineers or by exploiting the extracted information of a given model of the system. The latter technique is called model-based testing, and for complex software systems it is still an actively evolving field [52].

A typical verification plan usually starts by tackling corner cases with hand-written tests. The verification stimuli set is then improved by adding information automatically generated exploiting simulation-based approaches. At last, the automatically generated test set requires an additional analysis by the verification engineers. Tests developed in such a way require a considerable amount of expertise related to the device under test, they are not always portable, and their preparation is time-consuming and expensive.

Completely automated approaches for stimuli generation can follow several methodologies: constrained-random generation, sometimes simply referred to as random or pseudo-random test generation, and feedback-based generation are the most widely adopted.

In a constrained-random test generation [69], random stimuli set are created by following a constrained generation process. Templates and constraints previously specified are exploited to define the structure of each stimuli fragment which is then randomized. When targeting real designs, such techniques have been proved to be really challenging, and are outperformed by feedback-based approaches [114].

Feedback-based approaches initially apply stimuli to the system, check the output produced and obtain information that is eventually exploited to produce new, and probably better, stimuli. This process is repeated, generating a set of stimuli able to stress the system very effectively: considerable proofs support the predominance of feedback-based techniques over other simulation-based ones [145]. Another important advantage of feedback-based approaches is that at the end of the process, a very compact set of data is produced: even though a large number of stimuli is simulated, most of the results are fed back to the system and exploited internally. Thus, verification engineers are required to analyze smaller quantities of information.

In a typical hand-written test for a new mobile phone, the phone is woken up from deep sleep mode, a sequence of key pressures is given in input to it, it is turned back to deep sleep and power consumption is eventually determined. Frequently, these sequences of keys mimic actions that will be likely performed on the phone, e. g. starting a video call, inserting a new field in the address book, etc. Once a number of similar devised tests are completed, a test set is created thanks to an automated approach that generates stimuli similarly structured to the hand-written ones.

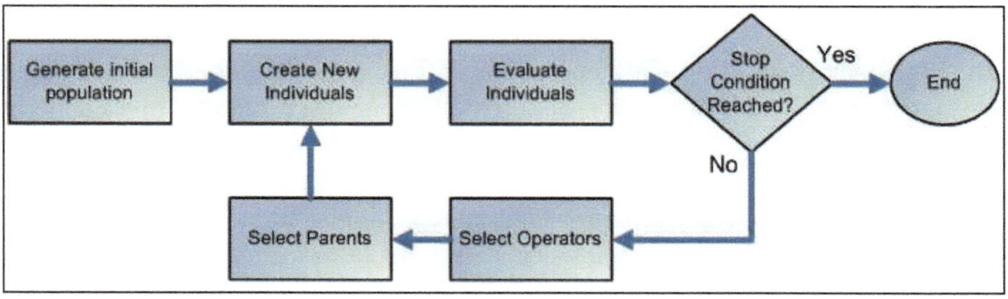

Fig. 3.2 Flowchart of a generic EA. During the evaluation step, individuals with lowest values of goodness are removed from the population.

Verification plans focused on simulation-based techniques are developed by industries to provide a set of stimuli able to excite completely the functionalities of the device under verification, consequently locating possible software bugs. When tackling the software of mobile phones, first of all verification engineers perform module-oriented verification procedures on single software application: this process is often developed separately for each component. In a second step, different applications are run at the same time, studying reciprocal influences among the modules and performing new verification tests on the whole system. In a third step, technical experts use the device, trying to locate weaknesses of the complete framework. During each step, verification engineers may rely on techniques available in literature on a single phase.

Among feedback-based techniques, Evolutionary Algorithms (EAs) are stochastic search techniques that mimic the metaphor of natural biological evolution to solve optimization problems [107]. Initially conceived at the end of 1960s, the term EAs now embraces genetic algorithms, evolutionary strategies, evolutionary programming, and genetic programming. Scientific literature reports several success stories in different domains, for instance [131].

Despite great differences, all EAs have many properties in common. EAs operate on a population of individuals; underlying each individual encodes a possible solution for the given problem. The goodness of every solution is expressed by a numeric value called fitness, usually obtained through an evaluator able to estimate how well the solution performs when applied to the problem. An evolutionary step, called generation, always consists of two phases: a stochastic one where some of the best individuals are chosen at random to generate new solutions; and a deterministic one, where solutions are ranked by their fitness and the worst ones are removed from the population. The process is then repeated until a user-defined stop condition is met. Fig. 3.2 shows a classical flow for an EA. When facing verification problems, stimuli created by an EA explore the solution space very efficiently. Moreover, the solutions found by EAs are somewhat very different from, and thus complementary to, human-made solutions [97].

3.3 Proposed Approach

The objective of the proposed approach is to find a set of stimuli able to detect errors triggered by the interaction of software applications on a mobile phone by stressing the functionalities of all the modules as much as possible. The approach is feedback-based, driven by an EA that evolves a population of candidate stimuli, coded as sequences of key pressures and pauses, similar to hand-written tests devised by expert engineers. The approach is also model-based: a finite-state machine (FSM) representing the system under verification is exploited to extract measures for the goodness of each solution. The FSM is automatically generated from scratch thanks to the data obtained by running simulations with the stimuli as an input to a physical prototype of the phone itself. The model supplies information on the number of different applications' features activated by each stimulus; this data is later used to assign a value to the stimulus, expressing its goodness. Fig. 3.3 shows a schema of the proposed framework: the EA manages a population of individuals that map stimuli. Such stimuli are evaluated by the model dynamically extracted from the physical device.

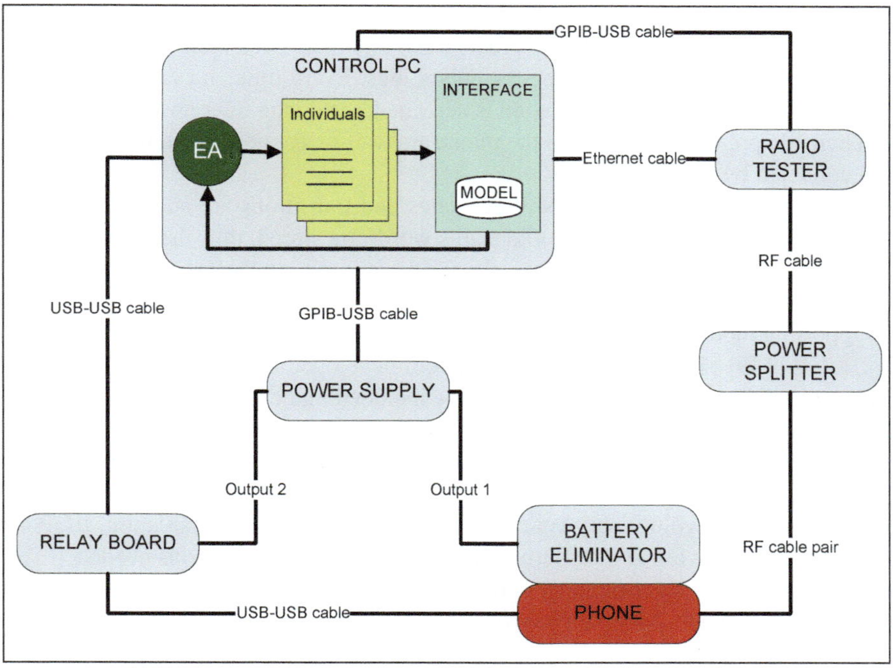

Fig. 3.3 Schema of the proposed framework

3.3.1 Model

The device under verification is modeled with a FSM, where each state defines a situation in which all active software modules are waiting for new inputs. A transition is a series of inputs that connect a state to another, turning on/arresting different applications or exciting some functionalities of the active ones. The FSM is exploited to evaluate the number of distinct states traversed and the transitions activated during the simulation of a stimulus [127].

Creating a complete model of all the software running on the mobile phone with the classical methodologies of software engineering would be impractical, requiring an excessive amount of time: the source code of each software module on the device should be provided and analyzed. Since some applications are developed by third parties, not all the software modules' source code is obtainable, thus critical data to build a complete model is missing. On the contrary, the FSM in the proposed framework is created as the simulations go on, and each time a new state is discovered the model is dynamically updated. Since this approach does not rely on a-priori knowledge, errors that could occur in the model-building phase are avoided.

The Operating System (OS) and most applications on mobile phones can run in a test mode where they write a log of their execution to ease the debugging process. By reading system messages recording applications starting and closing, called events in the following, it is possible to create a list of states. Each state is identified by a status word, obtained by parsing the debug logs. Every time an event is raised or a new feature of an active application is activated, the debug log register the changes. When all applications active on the phone are waiting for new input, the status word is collected by parsing the logs.

Starting with an empty FSM, new states and transitions are added each time a new status word is discovered. Old status words are stored, thus the framework can add transitions returning to states already known. Since the proposed framework makes mainly use of the number of different transitions fired, it does not require the supporting model to be complete or perfect.

μGP [149] [134], a general-purpose tool developed by the CAD Group of Politecnico di Torino, is the EA chosen to be included in the framework. μGP is available as a GPL tool [146]. Candidate solutions of a problem in μGP are represented as graphs, while the problem itself is encoded as an external program or script that evaluates each candidate solution and supplies the tool with a measure of its goodness. Since the evolutionary core is loosely coupled with the evaluation, μGP can be used in a wide range of different problems with no modifications needed.

While the tool was originally exploited to generate Turing-complete programs in assembly language, over the years μGP handled different problems whose solutions had complex structures.

Genetic operators, such as classical mutation and cross-over, modify the graph that encode the individuals. The tool architecture is designed to handle a large number of genetic operators, to ease the addition of new ones and to let the user choose the operators to apply to the problem. Each operator is associated with an activation probability, that is managed internally by μGP, and an endogenous parameter called *strength* that defines the differences between the parents chosen and the offspring generated.

In μGP version 3, individuals are represented as constrained tagged graphs, i. e. graphs with added information to nodes and edges, while the possible structures are limited by the user. Thanks to the constrained graphs, the tool can handle problems where the solution has structures simpler than Turing-complete assembly problems, like linear graphs, linear genomes or fixed-length bit strings.

The fitness of each candidate solution is computed by a script or program that runs a simulation using the individual as input and feeds back the results to μGP. The fitness in the tool is described by a vector of floating point numbers followed optionally by a comment. Each position of the vector is considered more important that the following: fitness A is greater than fitness B if the number in the nth position of vector A is greater than number in the nth position of vector B, and all the number in previous positions (if any) are equal; if all components are equal then the two fitness are considered equal.

The proposed framework makes use of μGP in its basic version, with no changes or additions to the original code. Configuration files in eXtensible Markup Language (XML) describe individuals' structure and all necessary parameters such as population size, stop conditions, number of genetic operators activated at each step. Since in the specific problem individuals map sequences of keys, the related graphs are linear genomes.

In the architecture of μGP, the evaluator is completely separated from the evolutionary core, so the evaluation program is designed from scratch and it is specific for each problem.

3.3.2 Candidate Solutions

Stimuli candidates to solve the problem are handled as a population of individuals by the EA. Each individual is a small program in Java that encodes sequences of keys and pauses: the programming language is chosen for the ease of compatibility with the OS running on the cell phone. The first part of each individual inscribes procedures of device initialization, while the last part makes the phone revert to an idle state to subsequently trigger deep sleep mode.

The initial population provided to the EA contains both individuals encoding random sequences of keys and pauses, and individuals encoding the most common actions performed by a user on the mobile device, e. g. selecting a number in the address book, making a video call, etc. Making the EA discover autonomously those sequences is possible, but it would take a great amount of time. Since human-devised tests have been already run on the prototype when the proposed methodology is used, starting the evolution from scratch is redundant. Individuals encoding common actions are derived from human-written tests cases used in other verification steps with an ad-hoc tool.

The EA manipulates and reassembles user-defined sequences and random individuals, mixing and modifying them to create new individuals with the aim to maximize the goodness of an individual. For example, a sequence derived from human-designed tests in the initial population may be later mixed with a different sequence and mutated by adding, removing or changing random lines of code.

3.3.3 Evaluator

Detecting a software bug that affects negatively battery life is the final goal of the evolution. Unlike other problems, where the goal leads straightforwardly to the definition of a continuous evaluation function, the presence of a bug cannot be expressed with such a function. A software error can be either detected or silent, with no other values. As natural evolution, EA "can act only by the preservation and accumulation of infinitesimally small inherited modifications, each profitable to the preserved being" [47]. The evaluation function of the specific problem needs consequently to be refined using heuristic methods.

The measure of the power consumption in deep sleep mode is surely included in the goodness value of each individual, because of the goal of the experience, but since most individuals use the same amount of energy, it is not enough to smoothen the landscape of the evaluation function. Parameters that lead to a quicker location of bugs must be taken into account as well.

The more an individual activates different software applications or different functionalities of the same application, the greater the probability that it will trigger a bug: consequently, individuals which excite more phone applications should be rewarded with a higher value when evaluating their goodness.

Three contributions (Pi, Ti, Ei) are taken into account for the global goodness value of individual i:

1. The mean value of power consumption while the cell phone prototype is in deep sleep mode, measured over 30 s and defined as

$$Pi = \frac{\sum_{t=0s}^{30s} P(t)}{30}$$

where $P(t)$ is the power consumption at time t;
2. The number of *transitions* covered in the FSM that models all the software applications running on the phone, as described in 3.3.1, defined as

$$Ti = \sum_{tr=0}^{TR} 1$$

where TR is the total number of transitions fired. A transition is defined as a passage from one state to another;
3. The number of different events activated, defined as

$$Ei = \sum_{e=0}^{E} 1$$

where E is the number of events raised from different applications.

The structure composed of these three contributes aims at discovering as many states as possible in the FSM built dynamically and at activating the maximum possible number of transitions. As in the initial idea, a high value is associated to solutions that excite a great number of different software modules on the phone, and have an extreme power drain in deep sleep mode.

As the measure of the goodness of the solutions is conceived, rewarding candidate solutions that activate more software applications could lead to the exclusion of an individual that targets only one module: that module, however, could be ignored by the rest of the population. In a similar way, the population in the long run could be filled with individuals very much alike. μGP, the EA chosen for the experience, has features enforcing diversity preservation in a population which help to avoid both those risks [43].

3.4 Experimental Results

The proposed framework was tested on a cell phone prototype running a Motorola P2K OS. The phone had been analyzed by verification engineers and passed all human-designed test. Thus, there were no known misbehaviors in the phone software modules when the proposed approach was applied to the device. The features of each component involved in the experience are summarized in Table 3.1. The experiments made use of the phone prototype, a radio tester, a power supply and a computer to control the instruments.

To measure the power consumption in deep sleep mode and to keep the phone powered during the experiments, a battery eliminator was connected to the phone. The battery eliminator is made of a battery whose contacts are disconnected from the inner cells and attached to a power supply. A relay board managed the connection between the PC and the device under verification: the phone does not enter deep sleep mode as long as it is connected to a PC. The relay board switched the phone from a state where it is in use to a state where it is no longer in use and can thus enter deep sleep.

To simulate a mobile network providing voice, video and data packet services, a radio analyzer was linked to the phone, thus producing an environment completely under the user's control.

By means of an ad-hoc tool, human-devised tests written in Java were converted to an XML representation later used as part of the initial population used by μGP. Such Java programs described every possible command that a user could issue to the mobile device, and each test was converted by μGP into a sequence of macro instances. The tool is problem-specific and it was developed in Motorola Research Laboratories located in Torino, Italy.

Since not all the applications stopped their execution when the mobile phone's flip was closed, the cell phone was sometimes blocked from entering deep sleep. For example, the software manager for the photo camera shows what the camera is shooting on the external monitor even if the flip is closed. Constraints were consequently modified to solve the issue. During the experience, every test was performed with a population of 50 individuals and an average of 40 new individuals generated at each generation. Each test took up to 100 generations of evolution.

With these parameters, it is clear that a great number of evaluations are performed during each experiment, so one of the first goals of the experience was to shorten the temporal length of a single evaluation as much as possible. Instead of setting the parameters of the phone by browsing through the menus, a time-expensive activity, by using seem elements, similar to configuration bits for the P2K OS, the duration of a test was reduced by 30 seconds. A time-out coded in the Java class that manages the logs of the phone was removed to further improved the performance by 20 seconds more. Nevertheless, some of the most dilatory steps could not be shortened: it took the phone roughly 60 seconds to enter deep sleep mode after the flip was closed, and the master clear needed to return the phone to its initial state after each test took 35 seconds.

Even with the improvements achieved, the average evaluation time of a candidate solution during the experiments was between 6 and 7 minutes, and thus it took about 5 hours to complete a single generation step. A significant saving of time was obtained thanks to the features of μGP, that keep the number of evaluations to a minimum.

To avoid the generation of multiple individuals triggering the same software errors, μGP constraints were altered after each discovery of a bug.

All the experiments had been completed in the Motorola Research Labs in Turin.

3.4.1 Video Recording Bug

During the first evolution, the majority of individuals showed a deep sleep power consumption between 2,5 and 3,2 mA. It took 16 hours of computation and 150 candidate solutions evaluated to find three individuals with a deep sleep consumption of about 7,0 mA which is 2,5 times the normal value.

The shortest of the three was composed by 100 lines of Java code, defining pressures of keys and pauses. It was analyzed through a series of runs on the framework already developed for the experiments, and the cause of the bug was uncovered: the pressure of specific buttons while the phone was in video recording mode caused a warning dialog to pop up on the display and froze the OS completely, thus keeping the phone from entering deep sleep. The error was caused by the interaction of the software controlling the video recording and the software managing the address book.

Further analyses on the two other candidate solutions uncovered the same subsequence of keys that caused the error in the first one.

3.4.2 Voice Call Bug

It took additional 120 hours of computation and the evaluation of about 1120 individuals to find out a second power-related bug. The best candidate solution found during this experience let the phone enter deep sleep, but the power measurements revealed a consumption of 50 mA, more than 16 times the expected use of power in deep sleep mode.

The mobile phone did not show messages or exhibit unexpected behaviors: it was necessary to analyze the individual's code line by line, but the cause of the error was eventually located. If the video call button was pressed along with a special series of keys during a voice call, the phone camera was powered up and it kept being active even when the device returned to deep sleep mode, thus consuming an extreme quantity of power. This error was triggered by an interaction between the software controlling the camera and the software controlling the voice call.

3.4.3 Incorrect Menu Behavior

The experiments uncovered a third misbehavior, not affecting battery life but not previously located by human-devised tests. By entering a specific settings menu and exiting hereupon without making modifications, the device reset some of its settings to their initial values. A candidate solution with this pattern made the following tests fail. The end user would probably not affected greatly by this misbehavior, but the problem had to be taken into account during the experience.

Table 3.1 Hardware involved in the experience

Device	Details	Description
Motorola mobile phone prototype	P2K Platform	Runs the tests
Personal computer	OS: Microsoft Windows XP, Primary Memory: 512 MB, Ports: USB 2.0	Controls the power supply, the radio tester and the phone; provides packet data services to the radio tester
Power supply	Double-output with measurement capabilities; VISA interface and SCPI protocol support	Controls the relay board; performs current drain measurement of the phone during tests
Anritsu MT8802A Radio Communication Analyzer	Model with the proper installed options (see below)	Simulates the cellular network providing voice, video and data packet services
Two VISA bus cables	Any supported VISA interface	Connect the control PC to the power supply and the radio tester
Power splitter	500 - 5000 MHz	Joins the signals of the two antennas of the phone (required only to test simultaneously GSM and 3G)
RF cable	N type connector (to the radio tester)	Connects the radio analyzer to the power splitter
Two short RF cables	QMA type connector (to the phone)	Connect the two antennas of the phone to the power splitter (actually only one cable is required if testing only WCDMA or GSM standard)
Battery eliminator	Built in house	Bypasses the phone battery, feeding the phone with the power provided by the power supply
USB relay board	Allows computer controlled switching	Switches on/off the USB connection between the phone and the PC

3.5 Conclusions

A framework for automated verification is proposed to attest the correct behavior of a cell phone, uncovering software defects not detected by human-devised tests, searching specifically for bugs affecting power consumption. The approach makes use of feedback from the device under verification to produce new stimuli, with an EA providing the necessary intelligence. The quantity of final data is limited to a small significant amount.

The effectiveness of the system is demonstrated on a mobile phone prototype implementing the P2K OS. The framework successfully locates software misbehaviors previously undetected by standard human-supervised verification. Two of the bugs uncovered are critical with regards to power consumption in deep sleep mode, thus making them high-priority from the user's point of view.

The research team at Motorola Research Laboratories in Torino finds a way to further improve the qualifying verification plan for mobile devices.

Chapter 4
Post-silicon Speed-Path Analysis in Modern Microprocessors through Genetic Programming

The incessant progress in manufacturing technology is posing new challenges to microprocessor designers. Nowadays, comprehensive verification of a chip can only be performed after tape-out, when the first silicon prototypes are available. Several activities that were originally supposed to be part of the pre-silicon design phase are migrating to this post-silicon time as well. This chapter describes a post-silicon methodology that can be exploited to devise functional failing tests. Such tests are essential to analyze and debug speed paths during verification, speed-stepping, and other critical activities. The proposed methodology is based on the Genetic Programming paradigm, and exploits a versatile toolkit named μGP. The chapter describes how an evolutionary algorithm can successfully tackle a significant and still open industrial problem. Moreover, it shows how to take into account complex hardware characteristics and architectural details of such complex devices. The experimental evaluation clearly demonstrate the potential of this line of research. Results of this work have been accepted for publication in [137].

4.1 Background

Nowadays, manufacturing technology is advancing at a faster pace than designing capability, posing unprecedented challenges in the arena of integrated circuits. The so-called *verification gap* denotes the inability to fully verify the correctness of devices that could be built, and indeed *are* actually built. Practice surpasses theory: comprehensive verification of a chip can only be performed after tape-out. Once manufacturing is completed and first silicon is produced, the early chips are sent back to their design teams. This process is called *post-silicon verification* to distinguish it from the traditional, pre-silicon, one. More generally, several activities that were originally supposed to be part of the pre-silicon design phase are nowadays migrating to the post-silicon time. The cost of manufacturing prototypical devices is enormous, but this practice is not an option. Designers candidly acknowledge

E. Sanchez et al.: Industrial Applications of Evolutionary Algorithms, ISRL 34, pp. 31–44.
springerlink.com

that "very few chips ever designed function or meet their performance goal the first time" [106].

Microprocessors are a paradigmatic example of the current trend: devices for the desktop market contain billions of transistors, implement complex architectures[1], and operate into the microwave frequency range. To give some examples, in a *pipelined architecture*, assembly instructions are executed as in a production line. Consequently, whereas the single instruction is not sped up, the global throughput is significantly increased. Even more, a *superscalar architecture* exploits duplicated functional units by executing two or more different instructions in parallel. The *branch prediction* unit guesses which way of a conditional branch will be taken, thus the execution may continue without waiting for the actual outcome of the test. Whether the conjecture was mistaken, a mechanism of *speculative execution* enables to efficiently roll back and undo changes.

Since last decade, desktop microprocessors also include hardware support to efficiently execute multiple *threads*, that is, independent flows of instructions. These architectures allow to increase the overall throughput in a multitasking environment, even when it would be impossible to further speed up the single program with the precedent techniques. *Simultaneous multithreading*[2] architectures enable multiple threads to be executed concurrently exploiting superscalar designs. More recently, in a *multicore architecture*, or *chip-level multiprocessor*, two or more independent processing units work side by side packaged in the same chip and sharing the same memory. Indeed, in modern multicore microprocessors each individual core also exploits simultaneous multithreading.

Besides this bewildering complexity, electric signals do propagate inside a microprocessor through different *paths*. To guarantee a correct behavior, all signals must reach a stable value within the current clock cycle, regardless the length or the complexity of their routes. It must be remembered that when a microprocessor is reported to operate at 3 GHz, the time available for signals to stabilize is slightly above 3 x 10-10 seconds. It may be hard to visualize such a frantic activity, for in this interval of time light covers only 10 cm (almost 4 inches).

Non-deterministic effects, such as manufacturing variability, are posing even greater challenges to the designers. It has been long known that several physical defects only appear when the device operates at full speed [152], but nowadays design criticalities also become apparent only at high frequencies. Even worse, they appear only occasionally, and possibly only in a percentage of the manufactured chips. "Finding the root cause of at-speed failures remains one of the biggest challenges in any high-performance design", stated Rob Aitken in his *editor's note* for [83].

---

[1] Some texts emphasize the difference between the specification of the machine language and its implementation, calling the former "instruction set architecture" and the latter "microarchitecture".

[2] Called "hyper-threading" in Intel designs.

4.2 Introduction

To meet today's performance requirements, the design flow of a modern microprocessor goes through several iterations of frequency pushes prior to final volume production. Such a process is called *speed stepping*. A *speed path* (or *speedpath*) is a path that limits the performance of a chip because a faster clock would cause an incorrect behavior. Speed paths may be the location where potential design fixes should be applied, and may indicate places where potential holes in the design methodologies exist.

At design time, the slowest logic path in a circuit is termed the *critical path*, and it can be easily determined. However, for complex high-performance designs, it has been recognized that critical paths reported from the pre-silicon timing analysis tools rarely correlate well to the actual speed paths. The reason is that any pre-silicon analysis tool is only as accurate as the model and the algorithms it uses. Obtaining 100% accurate process models for nanometer processes is difficult, if not nearly impossible. Analysis algorithms are also approximated because of the complexity involved. Moreover, timing behavior on the silicon is a result of several factors mingled together. But in the pre-silicon phase it would not be computationally feasible to consider all these factors simultaneously, and they are analyzed separately [164] [82] [25].

The identification of *failing tests*, i.e., sequences of operations that uncovers incorrect behaviors when run at high frequency, is highly related with speed path identification. Failing tests may be, for example, sequence of inputs to be applied to the microprocessor pins by an automatic test equipment (ATE). Such test are usually crafted with care by engineers starting from the pre-silicon verification test suite; generated by pre-silicon specialized tools, or automatic test pattern generators (ATPGs); or also created post silicon[3], tackling the actual devices [96] [165].

Interestingly, the instruction sets of microprocessors has been successfully exploited to tackle *path-delay faults*, i.e., manufacturing defects that slow down the signals covering a specific path inside the device [93] [35]. The underlying idea of these works is that executing a set of carefully designed programs may uncover timing issues. The main strength of the methodology is that the execution of such *test programs* is per se *at-speed* and requires no additional hardware, or complex and expensive ATEs. No attempts, however, have been reported to devise failing tests directly at the instruction level. No one has yet proposed a post-silicon methodology able to automatically generate a test program that stresses a speed path causing a detectable functional failure.

A *software-based speed-path failing test* is defined as an assembly-language program that produces the correct result only while the microprocessor operating frequency is below a certain threshold. As soon as the frequency is pushed above the threshold, the result yielded by the program becomes incorrect. Let us denote the threshold for a given program as its *functional frequency threshold*, because the incorrect behavior is functionally observable. That is, it can be

[3] The expressions "on silicon" and "silicon based" are also used.

theoretically detected without an ATE or other special equipment, simply by observing the values stored in the main memory and registers. Clearly, the diagnostic capability of a software-based speed-path failing test increases as its functional frequency threshold decreases. A test that produces a failure at a relatively low frequency is preferable to a test that fails only at very high frequencies.

This chapter shows how software-based speed-path failing tests with low functional frequency thresholds can be automatically generated by an evolutionary algorithm. Moreover, it demonstrates that the technologies already available in modern microprocessors can completely cut out the need of external equipments at the expense of a slight decrease in accuracy. The first result advocates for the exploitation of the methodology inside the manufacturer's facility during speed stepping phase. The second calls for coarse-grained, but quite inexpensive, incoming inspection campaigns.

Sections 3 and 4 describe the proposed methodology, detailing the adopted evolutionary algorithm. Sections 5 illustrates the feasibility study and report the obtained results. Section 6 concludes the chapter, sketching the future directions of the research.

4.3 Generation and Evaluation of Test Programs

The proposed approach for generating software-based speed-path failing tests is *pseudo-random* and *simulation-based*, or, more exactly, *feedback-based*. Candidate test programs are created without a rigid scheme, and evaluated on the target microprocessor. The data gathered are fed back to the generator and used to generate a new, enhanced set of candidate solutions. The process is then iterated.

To exploit such a mechanism it is indispensable to evaluate the goodness of each candidate test. As stated before, a software-based speed-path failing test is as good as it fails at low frequencies, and the key parameter in evaluating a test is its functional frequency threshold. However, it should not be forgotten that variability vexes verification engineers. A failing test may not fail always at the same frequency, even if all controllable parameters are exactly reproduced. The variability of speed paths may be caused by non-deterministic factors, such as noise, die temperature or small fluctuation in the external power. Some design criticalities may appear only under particularly unfavorable conditions. All experiments need to be repeated at least several times, when not on different devices.

Consequently, besides the lowest functional frequency threshold detected amongst the repeated experiments, an additional parameter in evaluating a test is the percentage of runs that actually failed at that frequency. It is intuitively plausible that a test failing half of the times at a certain frequency is more useful that a test that fails only every thousands experiments.

Changing the operating frequency of a microprocessor, however, is not an easy task. To ensure proper synchronization between all the components of the system, only a very limited set of operating clock speeds are available to the end users. While the microprocessor is connected to an ATE after production, such an evaluation is perfectly feasible. However, outside manufacturer laboratories the large steps in frequencies would likely impair the overall usability. Notably, outside manufacturer laboratories, the final aim would hardly be speed stepping. Conversely, end users may be quite interested in performing an incoming inspection on purchased devices. Tacking this latter goal, this chapter shows how to adapt the methodology in order to require no test equipment and no additional hardware whatsoever.

The architecture of modern microprocessors includes dynamic performance scaling technologies. Intel branded it as *SpeedStep*. Similar mechanisms are available as Advanced Micro Devices *PowerNow!* and *Cool'n'Quiet*, or VIA Technologies *LongHaul*. Such technologies are designed to save power and reduce heat, thus they allow to decrease the operating frequency and the power supply voltage supplied to the microprocessor. Reducing the CPU core voltage is known as *undervolting*.

Roughly speaking, desktop microprocessors are made using the complementary metal-oxide-semiconductor (CMOS) technology, based on field-effect transistors (FETs). In such devices, reducing the voltage increases the time required to switch between logic values [14]. Thus, the effects of reducing voltage may be reasonable related to the effects of increasing the operating frequency. As a matter of fact, whenever a microprocessor is undervolted, its operating frequency is also reduced to guarantee proper functionalities. Manufactures define sets of safe operating states, sometime called *performance states* or *p-states*. While the exact meaning of these p-states is implementation dependent, P0 is always the highest-performance state, with the following *P1* to *Pn* being successively lower-performance and less-consuming states.

Following the discussion, it appears evident that undervolting a microprocessor emphasizes speed-path criticalities. Moreover, reducing the core voltage cannot damage a device. Thus, to stress speed paths the behavior of a microprocessor could be analyzed intentionally outsides the predetermined p-states. Let us define the *functional core voltage* of a failing test as the lower voltage required not to fail the test at a given operating frequency. Conversely to functional core frequency, a failing test is as good as its functional core voltage is high. That is, all tests would fail with a very low core voltage, but only the interesting ones truly require full power.

Thus, an alternative evaluation of a candidate test could be based on its functional core voltage, and on the percentage of runs that actually failed.

4.4 Evolutionary Approach

The proposed test-program generator exploits a versatile evolutionary toolkit called μGP developed at Politecnico di Torino, and available under the *GNU Public License* from *Sourceforge* [146]. Unlike usual genetic programming (GP) implementations, μGP specific target is to produce realistic assembly-language programs. Its original purpose was to assist designers in the generation of programs for the test and verification of different microprocessors, hence, the Greek letter micro in its name.

μGP was designed to support assembly peculiarities, like various conditional branches, different addressing modes, or instruction asymmetries. Generated programs take advantage of syntactic structures as global and local variables, subroutines and interrupts. Since its creation, the tool underwent three main revisions [40], [149] and [134]. The latest version internally encodes individuals as directed multigraphs, and this enable the handling of a quite wide range of problems.

μGP is asked to devise an assembly program to be used as a software-based speed-path failing test. Following the previous discussion, a population of candidate test programs is evolved, and the evaluation of their goodness is used as fitness function to drive the process. However, the specificity of the task calls for several different problem-specific knacks.

4.4.1 Fitness Function

During experiment the system frequency is first increased using the so-called overclocking features of modern main boards. An excessive increase of the frequency may cause overheating or otherwise irreparably damage the microprocessor, but increasing it slightly is usually perfectly safe. Then the evaluation is performed by reducing the core voltage, only.

Similarly to *software-based self test* [132], candidate test programs include a mechanism that help checking their own correctness: all the results of the calculations performed by the test program are compacted in a single signature using a hash function. The evaluator runs the test program in safe conditions, i.e., at full power, and store the signature. Then it runs the program again at decreasing CPU core voltages, checking that the signature is not modified. As soon a difference is detected, the functional voltage threshold is recorded. The whole process is repeated R times to tackle variability.

In μGP the fitness function may be specified as a vector of positive numbers. The components of the vector are strictly hierarchical, with the first being the most important. The first component of the fitness value is simply the functional voltage threshold. The second is the number of failures detected over the R repetitions at the maximum voltage. It must be stressed out that the actual result of the calculations is of no interest, the only relevant detail is that it changes when the test is executed undervolting the CPU below the functional core voltage.

4.4.2 Individual Evaluation

μGP creates assembly functions, that are assembled and linked with a *manager* module. These functions contain a loop that execute L times a set of instructions.

The instructions themselves are devised by the evolutionary core, while the framework is fixed. At the end of the loop, before the next iteration, the values in the registers are used to update the signature.

In the proposed methodology the very same microprocessor is used both for generating candidate tests, i.e., for running μGP , and for their evaluation. Using the same processing unit to evolve individuals and calculate their fitness is quite a standard procedure in GP. In most μGP application, conversely, the interesting data is not the *result* of the computation, but *how* the test program is actually computed by the specific device. And the evaluation of the assembly-language test programs is usually carried out on an different unit, physically, by emulation, or by simulation. Extracting information from the microprocessor currently executing μGP may be quite tricky. It has been first attempted 2004, during a collaboration with Intel [97].

When it is required to calculate the fitness of the newly generated offspring, individuals are compiled to stand-alone executable and run. The manager also takes care of invoking the evolved fragment of code while varying the CPU core voltage, and creating a text file with the results. Eventually, the execution of μGP is resumed.

4.4.3 Evolution Start

Evolution advances *through the accumulation of slight but useful variations* [47]. Thus, if all individuals in the initial population are indistinguishable, it is hard for the process to start. Unfortunately, this is not an uncommon situation. The computer used for generating and evaluating the test programs is almost completely working. It is able to perform nearly all operations, and indeed finding an incorrect behavior requires elaborate sequences of instructions. Thus, in the first step it is not infrequent to have a population of test programs not able to fail at any voltage, with exactly the same fitness value.

To overcome this problem the first population is significantly larger than the usual ones. μGP uses the parameter ν (the Greek letter nu) to control the number of randomly generated individual in the beginning of the evolution.

4.4.4 Internal Representation, Multithreading and Multicore

In μGP , the individual is internally encoded as a directed multigraph. With the adopted scheme, disregarding all the details, each node encodes a line of the assembly program. Edges represent syntactic or semantic relationship. For instance one edge connects every two adjacent lines; an additional edge connects a branch instruction with its target; another edge connects a node referring to a global variable with the line defining the data.

Modern processors may implement a multithreaded design; or they can exploit a multicore architecture; or even both. From the perspective of the test-program generator details are not relevant, but it is vital to create multiple independent instruction flows.

A single individual is composed of different independent functions. The manager activates them as different threads on different cores using appropriate operating system calls, or directly whether no operating system is used. Such blocks, in the individual, are represented as disjoint subgraphs. Notably, different blocks may be forced to have different structural characteristics, or use different subsets of instructions.

4.4.5 Assembly Language

For the generation of failing test is performed during speed stepping or an incoming inspection, it is essential to test all possible instructions, and especially the newest. The assembly instructions made available to μGP can be divide in three main classes.

Integer instructions include all usual instructions, such as logical and arithmetical ones. They operate on internal registers or memory. In the adopted scheme, only two registers are employable, while the others are used by the manager. However, this restriction should not impair the global result. Comparisons, tests and branches are also included in this class. To avoid endless loops, μGP was forced to create only forward branches in the generated code.

x87 instructions are the subset of the Intel 32-bit architecture (IA32) related to the floating point unit (FPU). The name stems from the old separate floating point coprocessors, like 80287 and 80387. They provides single precision, double precision and 80-bit double-extended precision binary floating-point arithmetic according to the IEEE 754-1985 standard. x87 instructions operates on a stack of eight 80-bit wide registers, but some instruction modifiers allow the use of the stack as a set of registers. In the actual version, μGP uses x87 instructions in only one thread.

The third class of instructions requires a slightly longer introduction. In 1996, Intel introduced *single-instruction/multiple-data* (SIMD) *instructions* in the *Pentium* microprocessor, its first superscalar implementation of the x86 instruction set architecture. In a SIMD instruction, multiple processing elements perform the very same operation simultaneously on different data. Matter-of-factly, the technique is called *data-level parallelism*. Pentium SIMD instructions were originally branded as *MMX extension*, and operate on eight 64-bit wide registers. Advanced Micro Devices offered its own enhanced version of the SIMD instructions two years later, marketing them as *3DNow!*. In 1999, Intel outbid with the so-called Streaming *SIMD Extensions*, or *SSE*. Followed in 2001 by *SSE2*, in 2004 by *SSE3*, and finally in 2006 by *SSE4*. Not mentioning the *Supplemental Streaming SIMD Extensions 3* (SSSE3, with three “S”) included in Intel microprocessors from 2006. Advanced Micro Devices is planning to include *SSE5* in its *Bulldozer* processor core in 2011.

Not surprisingly, SIMD instructions are particularly critical during speed stepping. The complex calculations involved by these instructions cause data to go through several functional units, and the resulting *datapaths* are prone to be source of problems when the operating frequency is increased.

4.4.6 Cache

Cache memories are small, expensive and fast memories placed near the processor core. The rationale is to read and write the most frequently accessed data as efficiently as possible. Modern microprocessors exploit a hierarchy of cache memories, or *multi-level caches*, with the level-1 (L1) cache being the smallest, more expensive and fastest. And, indeed, the closest to the central processing unit.

When the memory is accessed, the L1 cache checks whether the data is *cached*, i.e., if it contains the specified location. In this case, called *cache hit*, the L1 swiftly replies to the request. If the data is not present, termed *cache miss*, the L1 cache delivers the request to the L2 cache and so on. Considering only the first level, there is a significant difference in performance and power consumption between a L1 cache hit and a L1 cache miss. Such effects may be significant for the generation of a failing test, and must be taken into account.

The internal design of a cache is complex, and the policies for determining which data to store and which to discard are different. In a *fully-associative* cache, every memory location may be cached in every location of the cache. However, such a design is too complex and slow if the size of the cache increases. Thus, usually, the architecture imposes that a specific memory location may be stored only in a subset of cache locations. In a *direct-mapped cache* each memory location can be cached in only one location, while in a *k-way set associative cache*, in k alternative locations.

In order to give the μGP the possibility to generate cache hits and cache misses, a special set of *C* variables was defined. The variables are carefully spaced so that all their memory locations will be cached in the very same cache location. If the microprocessor uses a k-way set associative L1 cache and `C > k`, a shrewd sequence of read and write operations on such variables may generate the desired cache activity.

It must be noted that the goal of adding such variables is to let the evolutionary core to control the cache activity, but no suggestions are given on how to exploit them. μGP would devise which sequence of operations is more useful to generate a failing test.

4.5 Experimental Evaluation

While no working attempts of functional failing-test generation has been reported in the specialized literature, a related problem is faced by a community of computer enthusiasts. *Overclockers* try to push the performance by increasing the operating frequencies of their microprocessors and the CPU core voltages [38]. However, after pushing their computers to astonishing frequencies, they need to assess the stability of their systems. The test suites that are used to stress the systems and highlight criticalities may be regarded as generic fail tests not focused on a specific microprocessor. Thus, they can be used as a baseline to evaluate the performances of the proposed methodology.

While all the stability tests are quite different, a common point is that modern ones do extensive SIMD calculation. Another common point is their ability to increase the temperature of the microprocessor. It is well known that high temperature may cause both reversible and irreversible effects on electronic devices. Heating may increase the skew of the clock net and alter hold/setup constraints, causing design criticalities to become manifest and the circuit to operate incorrectly [27].

However, while such an effect is sensible when assessing the stability of a system, it may not be desirable when the goal is to find a failing test during speed stepping. The main reason is that the failing test should be as repeatable as possible, while increasing the temperature also increase non-deterministic phenomena. Nevertheless, since no other comparison is possible, the proposed approach was tested against the state-of-the-art stress tests used by the overclocking community.

4.5.1 Overclockers' Stress Tests

Most of the information about stability stress tests is available through forums and web sites on the internet, with few or none official sources. However, there is quite a generalized agreement in the overclockers community on these tools.

SuperPI is a version of the program used by Yasumasa Kanada in 1995 to compute π to 232 digits. It is based on the Gauss-Legendre algorithm. SuperPI implementation makes use of x87 instructions only, it exploits no SIMD instructions, and it is strictly single threaded. CPU BurnIn is a stress test developed by Michal Mienik in the beginning of 2000s. Like SuperPI it uses no SIMD instructions and is single threaded. These two programs are rather old, but have been included for the sake of comparison.

Prime95 is the name of an application written by George Woltman and used by a project for finding *Mersenne prime numbers*[4] [1]. It makes extensive use of the fast Fourier transform, or FFT, with a highly efficient implementation that exploits SIMD instructions. Over the years, it has become extremely popular among overclockers as a stability test. It includes a "Torture Test" mode designed specifically for testing systems and highlight problems. In the overclocking community, the rule of thumb is to run it for some tens of hours.

LINPACK is a software library for performing numerical linear algebra on digital computers. It was originally written in Fortran in the 1970s and early 1980s.

[4] A *Mersenne number* is a positive integer that is one less than a power of two: $M = 2^p - 1$. The name came from the French theologian, philosopher, mathematician and music theorist Marin Mersenne, sometimes referred to as the "father of acoustics". As of August 2010, only 47 Mersenne prime numbers are known. Remarkably, the largest known prime number is also a Mersenne number: $N = 2^{43,112,609} - 1$.

Newer implementation of LINPACK exploits SIMD and are highly optimized. Significantly, Intel includes a benchmark based on an optimized version of LINPACK in its Math Kernel Library [2]. Different applications exploited such benchmark to assess the stability. The most common are *LinX*[5], *IntelBurnTest*[6], and *OCCT*[7]. The last one, also includes a proprietary stress test.

4.5.2 Target System

Experiments were run on an Intel Pentium Core 2 Duo E2180, MSI motherboard NEO2-FR with the Intel chipset P35. The system was equipped with 3 GiB RAM memory DDR2-800, and a Sparkle Nvidia 8800GT graphic card. While the default clock was 2GHz, for the purpose of the experiments the system was overclocked to 2.93GHz. The only non-standard device was an in-house manufactured water cooling system (Fig. 4.1).

The E2180 is a dual-core microprocessor. It has a 32 KiB L1 cache for data implementing an 8-way set associative architecture. An identical cache is for instructions. The L2 cache is 1 MiB, 4-way set associative, and it is used for both data and instructions. The *Core* architecture can be traced back to the *P6*, introduced in 1995 with the *Pentium PRO* and revived in 2000 with the *Pentium M* line. It supports SIMD instructions up to SSE3 and SSSE3, and the *Enhanced Intel SpeedStep* (EIST) technology. Unlike its predecessor *NetBurst* and its successor *Nehalem*, the Core 2 Duo architecture does not exploit simultaneous multithreading.

Given the goal of the feasibility study, the difference between multicore and multithread may be regarded as a marginal detail. From the perspective of μGP there is no difference whether the different threads are evaluated on the same core or on multiple cores.

4.5.3 Experimental Results

The failing test devised by the proposed approach on the target system was compared with the state-of-the-art stress tools used by overclocking community. Results are reported in Table 4.2 and Table 4.3. Columns are labeled with the name of the program used to test the system. The last column reports data of the test generated by μGP. Rows indicate the CPU core voltage at which the experiments were run. Cells shows the time required for the given stress test to report a failure. To reduce overheating effects, all tests were stopped after 10 minutes. Thus "more than

[5] Originally posted on `http://forums.overclockers.ru/`

[6] `http://www.ultimate-filez.com/`

[7] `http://www.ocbase.com/perestroika_en/`

Fig. 4.1 The system used for the experiments.

10 minutes" means that no failure has been detected. All experiments have been repeated 10 times. μGP parameters are shown in Table 4.1.

Table 4.1 μGP parameters

Parameter	Meaning	Value
μ	Size of the population	30
ν	Size of the initial (random) population	100
λ	Genetic operators applied in each generation	20
R	Repetitions of each test to tackle variability	10
L	Repetitions inside each test	5,000,000

Table 4.2 compares the proposed methodology with older stress tests. Since multiple threads are not supported by SuperPI and CPU BurnIn they were disabled in μGP as well. It can be noted that the critical functional voltages are quite low, thus the microprocessor needs to be undervolted significantly in order to originate a

problem. Table 4.3, on the other side, reports the comparison against newer stress tests. All these programs uses two threads, that is, one for each core.

Table 4.2 Failing-test duration for single thread

CORE V	SuperPI	CPU BurnIn	μGP
1.2625	...	5′	1″
1.2750	10′	> 10′	1″
1.2875	> 10′	> 10′	...
1.3000	> 10′	> 10′	...
1.3125	> 10′	> 10′	...
1.3250	> 10′	> 10′	...

Failing tests devised with the proposed methodology clearly outperform all the other approaches. However, it must be noted that the comparison is not completely fair, since the goal of the programs were different. μGP was asked to find a very fast failing test for a specific microprocessor, and there is no guarantee that they would fail on different models. Moreover, the test was required to be very short, to avoid heating effects. On the contrary, the adopted stress tests intentionally exploit overheating and are designed to work with different architectures.

Table 4.3 Failing-test duration for multiple threads

CORE V	Prime95	IntelBurnTest	LinX	OCCT	μGP
1.2625					2″
1.2750					2″
1.2875	4′			7′	2″
1.3000	>10′	7′	7′	>10′	10″
1.3125	>10′	>10′	>10′	>10′	8′
1.3250	>10′	>10′	>10′	>10′	

The final failing test is 614 line long. The two functions executed by the two cores are respectively 280 and 235 line long. The remaining lines are mainly used to define and initialize variables or other program parts. It should also be noted that μGP requires about 50’ to generate a test failing at a core voltage of 1.2625V; 6h to find a test failing at a core voltage of 1.2750V; additional 5h for 1.2875V; and additional 5h for the 1.3000V. For the sake of experimentation, the failing test devised for 1.3000V was run at a core voltage of 1.3125V and consistently failed in about 8’. Interestingly, the temperature of the microprocessor during this last experiments never exceeded 40°C, while running LINPACK-based stress tests it is permanently above 45°C.

4.6 Conclusions and Future Works

An efficient post-silicon methodology for devising functional failing tests is proposed. The result of the chapter is twofold: first, it demonstrates the possibility for an evolutionary algorithm to generate assembly-level failing tests, tackling the most advanced microprocessor designs; second, it shows that the methodology can produce interesting results with negligible, or even nil, hardware overhead.

The proposed methodology could be exploited by microprocessor manufacturers, during verification or speed stepping. Or it could be used to generate a fast test able to check the reliability of a system. The latter can be important for the incoming inspection of a set of purchased devices.

Future works include enhancing the evolutionary algorithm, letting it tuning the number of repetitions in each test L. The interaction between x87 and SIMD instructions also deserves a closer examination. A customized version of the μGP requiring no operating systems can be devised in order to more easily run experiments on the microprocessor. Also, the signature could be improved by including more information on the state of the execution, such as the internal performance monitor.

Part II
Design and Reliability Problems

Chapter 5
Antenna Array Synthesis with Evolutionary Algorithms

This chapter describes an evolutionary approach to the optimization of element antenna arrays. Classic manual or automatic optimization methods do not always yield satisfactory results, being either too labour-intensive or unsuitable for some specific class of problems. The advantage of using an evolutionary approach is twofold: on the one hand it does not introduce any arbitrary assumptions about what kind of solution shows the best promise; on the other hand, being intrinsically non-deterministic, it allows the whole process to be repeated in search of better solutions. A generic evolutionary tool originally developed for a totally different application area, namely test program generation for microprocessors, is employed for the optimization process. The results show both the versatility of the tool (it is able to autonomously choose the number of array elements) and the validity of the evolutionary approach for this specific problem.

The experience described in this chapter has been presented in [101].

5.1 Introduction

Antenna arrays have long been used to achieve performance impossible to obtain from a single antenna. High-directivity antennas and shaped beam arrays are examples of products that take advantage of the array concept. Uniform arrays, however, may be unsuitable for a given specification. This drives us to the need for array synthesis and optimization, in order to obtain a given functional specification at a reduced cost. Numerous manual or automatic methods exist to achieve this goal: Conjugate Gradient [34], Fourier series and Woodward- Lawson methods [53] first explored the concept of automatic array synthesis; Monte Carlo method follow as a statistical approach [130] and finally genetic algorithms are used.

Previous work in this field includes the use of GAs [102], evolutionary programming [29] and hybrid methods [76].

Marcano and Duran [102] introduce the use of GAs for the optimization of linear and planar arrays. However, the problems presented do not seem to be

E. Sanchez et al.: Industrial Applications of Evolutionary Algorithms, ISRL 34, pp. 47–54.
springerlink.com

particularly stressful to the method employed. Chellapilla and Hoorfar [29] present an EP method for the generation of optimally thinned linear arrays, showing increased performance with respect to GAs. Hoorfar and Zhu [76], finally, show that hybrid methods perform better than pure GA or EP algorithms on some problems.

A rather generic evolutionary tool was used to address the problem of array synthesis and optimization. One peculiarity of the proposed approach is indeed the use of a tool developed for a totally different application area, namely test program generation for microprocessors. This not only allows us to critically assess the validity of the evolutionary approach to array synthesis, but also helps the development of the tool itself. Some of its new features, in fact, have been added on the consideration of their usefulness for this specific application, and are also being used in the original context.It is interesting to note that the used tool shows hybrid GA/EP properties since it employs both mutation and crossover. The tool itself will be described later. The chapter is organized as follows: a brief introduction on antenna arrays is given in section 2; section 3 introduces the evolutionary computation paradigm and describes in more detail the evolutionary tool used; in section 4 the workflow and the performed numerical experiments are described; section 5 reports the obtained results; finally, the conclusions are reported in section 6.

5.2 Antenna Arrays

particular actuators and sensors: the antennas. High gain applications require high directivity antennas; this can be achieved by arranging them in an array: more antennas are placed near each other to fuse their individual irradiation diagrams to obtain a collective diagram more fitted to specified application. Also it is possible to design antenna arrays with a shaped beam; these arrays irradiate in a particular space zone according to a pre-arranged form (for example: for a satellite which must irradiate a country one must design an antenna that has a shaped beam which covers only the desired territory). However the design of this type of antennas presents, unfortunately, various problems.

The problems which one meets during the design of a shaped beam antenna are substantially due to the fact that the design operation is of inverse type: from the normalized array factor that must be passed, to the position of radiators and to their feeding phase. To represent the array factor rigorously it is possible to express it as a polynomial whose roots represent the feed coefficients of the radiators. Changing the modulus or the phase of a root changes the overall shape. Another important problem is that the various radiators must furthermore be in such positions that their mutual coupling be minimum. With all these constraints the problem becomes quickly intractable. Also in the past the problem was relegated to the most expert designers; they started with different mathematical methods to do the synthesis of the antenna and, with little shifting of the various radiators, were able to obtain good approximate results; however the cost in terms of time was huge. The development of the computer technology gives us, today, various methods with which it is possible to automatically design this type of antennas, and with good results.

In the past the growth of the antennas was of evolutionary type, from the first systems to the more sophisticated ones. For example, going from the first ground plane antennas which presents an impedance of 38 Ω, passing on to the ground plane with folded arms with a 50 Ω impedance, to finish with the skirt dipole. Imitating this process, it is possible to get working and well approximated solutions, beginning from inefficient ones, with an evolutionary method.

5.3 Evolutionary Algorithm

Evolutionary computation is a computer-based problem solving paradigm based on Darwin' evolution theory [9]. In this paradigm possible solutions to a given problem are seen either as individuals inside a larger population or as species within an environment. These compete against each other and periodically undergo a selection process. The best solutions, i.e. the 'fittest' ones, survive the selection and are allowed to reproduce, that is to produce other solutions similar, but not completely identical, to themselves. These offspring are in turn subjected to the same selection process as their ancestors. This process leads, in turn, to an increment in the average fitness. The term fitness is historically used to denote a measure of the compliance of a candidate solution with its goals. An increment in the average fitness usually goes together with an increase in its maximum value. Evolutionary computation itself has evolved over time, producing many different kinds of evolutionary algorithms. The best-known ones are Genetic Algorithms, Evolutionary Programming, Evolution Strategies, Classifier Systems and Genetic Programming. None of these methods is perfect for all problems, but they offer a large choice of approaches for the user to try. Evolutionary methods are particularly suited to solve computationally hard problems for which no good heuristic is known.

The main goal of an evolutionary method is to make a computer obtain an exact or, more often, approximate solution to a problem without being explicitly told how to do so.

In the proposed approach, a tool named μGP is used. μGP [134] is an evolutionary approach to generic optimization problems with a focus on the generation of test programs for microprocessors, similar to both Evolutionary Programming and Evolution Strategies. It is not strictly a genetic algorithm since it does not employ a fixed-size chromosome setting, but a graph structure, to describe the individuals it cultivates. In Evolutionary Algorithms parlance, it is a steady-state evolutionary method that implements a variation of the $(\mu+\lambda)$ strategy on a single population of individuals. This means that, given an initial population of μ individuals, λ genetic operators are applied on it to produce a variable number of offspring; the parents and offspring are then merged into a single population, which undergoes selection: the μ individuals with the highest fitness are selected for survival, and the rest are discarded. Individuals with high fitness may remain indefinitely in the population.

It is different from Evolutionary Programming mainly because it employs crossover, currently in two forms; additionally, mutation operators are not implemented in many forms for strength selection, but rather a great variety of operators is implemented. Additionally, population selection is always deterministic. In common with Evolutionary Programming there is no requirement that a single offspring be generated from each parent. It is also different from Evolution Strategies in that it (currently) only employs the $(\mu + \lambda)$ strategy. It is, however, conceptually similar since its evolutionary basis is the individual, not the species. It finally differs from both since it dynamically self-adapts many of its parameters.

One of the main peculiarities of μGP is the fact that the focus during the reproduction process is not so much on the reproducing individual as on the genetic operator employed. In fact, the λ in the $(\mu + \lambda)$ expression is not the number of generated offspring as the standard terminology dictates, but the number of genetic operators used.

Although its original focus is the generation of test programs, μGP is a very versatile tool that can be employed to successfully approach a number of other problems, on the only condition that a solution can be expressed with the syntactical constraints as an assembly program. So, for example, any problem whose solution can be represented with a table, a tree, or a directed graph is eligible for approach.

The evolutionary core is continually being developed, and many features have been added to it over time, many of which may seem somewhat odd, to improve its performance: clone detection and optional extermination to avoid the evaluation of identical individuals and to improve genetic variety; a fitness hole in tournament selection, that is a small but nonzero probability that the tournament selection criterion is not the fitness but the entropy value of the individuals, again to improve genetic variability; parallel fitness evaluations; an initial population size optionally greater than μ, to better exploit the initial random search phase. In this chapter the support for real numbers in the individuals and a new form of mutation for μGP have been developed, and new features can be expected to appear in the near future.

5.4 Experimental Setup

The main goal of the numerical experiments was to obtain a working environment through which an automatic process of array synthesis and optimization could be performed. One of the main objectives is to reduce as much as possible the manual effort of the human designer, while still obtaining an acceptable solution. To set up the environment a very simple instruction library was implemented for μGP, specifying the allowed range for the roots of the array factor. The only thing the designer is left to do is specifying a wanted array factor, and optionally a desired maximum number of elements.

In the experimental setting, μGP is used to minimize a measure of distance between an objective array factor and the synthesized antenna's own array factor. Three different measures of distance are employed to evaluate the effect of various criteria on the quality of the result. The objective array factor is passed directly to the fitness evaluator, in the form of a series of $(\psi, F(\psi))$ values. The evaluator reads and normalizes this series, builds a second series containing the corresponding values of the current array factor, with the same normalization, then computes one of the three distances between the two series, as configured in a parameter file. The distances implemented so far are the classic sum of absolute differences, root-mean-square and maximum absolute difference between the two series.

The obtained results are rather different from each other, as will be shown in the next section. This reflects the importance of a careful selection of the fitness function.

As the parameter that most critically influences the quality of the solution found by an evolutionary method is the population size, very big populations are used in the proposed experiments. Also, a less hard selection scheme was employed , to let the evolutionary core explore a greater portion of the search space. To test the suitability of the approach, two types of experiments were performed: in the first one the objective was to approximate the array factor of an uniform array, while the second concerned the synthesis of a rectangular array factor. Approximating the uniform array is seemingly trivial, but, since the evolutionary tool starts from random solutions, it is not granted that it will quickly converge to the exact solution. The approximation of a well-known array type, moreover, gives us confidence in the employed methodology and lets us assess the quality of the obtained solutions. The rectangular array factor, on the other hand, allows us to push the method used to its limits, evidences the differences in performance between the various measures of distance and provides us further insight on the best ways to improve the fitness evaluator. While performing the optimization, it become noticeable that the choice of the initial number of roots has a noticeable effect on the achieved quality of the solution. This is due to the fact that the initial phase of the evolutionary method consists of a random search: giving the right number of roots allows the algorithm to randomly hit promising regions of the search space that would remain hidden during a normal search process that starts from a low number of roots. In this latter case, in fact, the evolutionary algorithm may generate solutions with the right number of roots when it is already in the exploitation phase, with a very uniform population, and thus unable to broadly explore the resulting higher-dimensional search space. Only the most significant results are therefore provided.

5.5 Experimental Results

The optimization on the uniform array approximation was conducted with a population of 300 individuals, applying 200 genetic operators per generation and carried on for 100 generations. The obtained results clearly show that even the approximation of an array factor is not a trivial operation. The best fit is obtained using the root-mean-square measure of difference between the objective function and the approximating function. The sum of absolute values yields a somewhat worse performance since it does not discriminate between small and large deviations from the objective, but lumps everything together with the final sum. The worst result of all is obtained using the maximum absolute difference between the two functions; this happens because the fitness landscape has large flat regions in it. To give a hint of why it is so, consider an objective function and a given candidate solution that has a specific value in point ψ_0 (named F_C), where the difference between F_C and the corresponding value of the objective function (named F_O) is maximum; call M this maximum; there obviously exists an infinite number of functions that pass the (ψ_0, F_C) and remain within distance M from the objective function and therefore exhibit the same fitness as the first one. This makes it extremely difficult to find a path even to local maxima. Figure 5.1 shows the results obtained with the three fitness measures. For the case of the rectangular objective function, a large population of 3000 individuals is used, applying 2000 genetic operators per generation and allowing the evolution to proceed for 1000 generations. The rectangular array factor proves a much harder problem to solve than the uniform array factor, not only needing more elements for an acceptable approximation, but also showing a poorer quality of the solution (Figure 5.2).

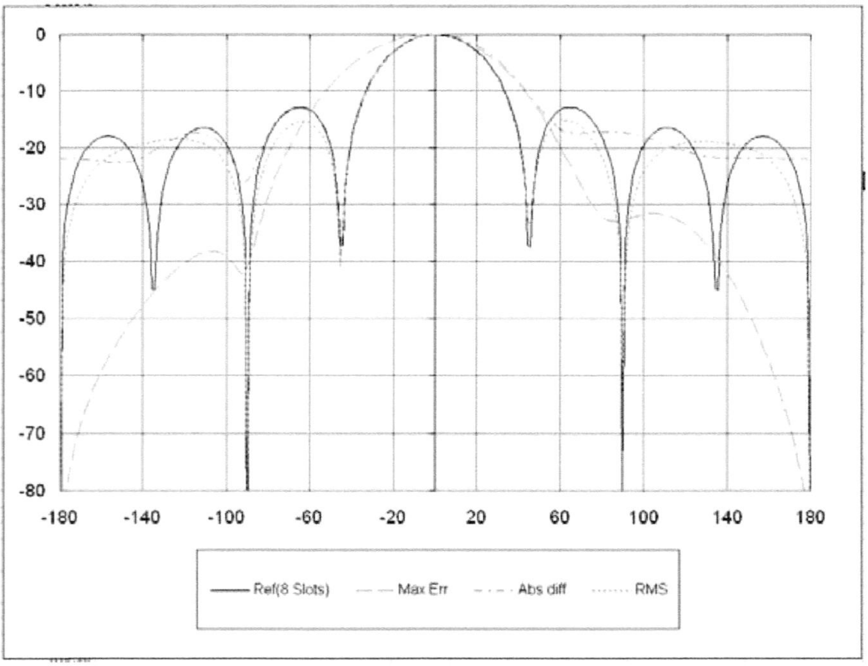

Fig. 5.1 Approximations of the uniform array factor

For a comparison, a similar numerical experiment performed approximating the uniform array leads to a result visually indistinguishable from the objective. Again the performance of the three fitness measures shows the same order. The root-mean-square difference measure leads to an imperfect approximation of the low level of the objective function, but to the overall better approximation of the high level; the sum of absolute differences yields the best approximation for the low level but a slightly worse aproximation of the high level; finally, for the same reasons outlined above, the maximum absolute difference gives us the worst performance, and the resulting evolutionary process is unable to satisfactorily approximate the objective.

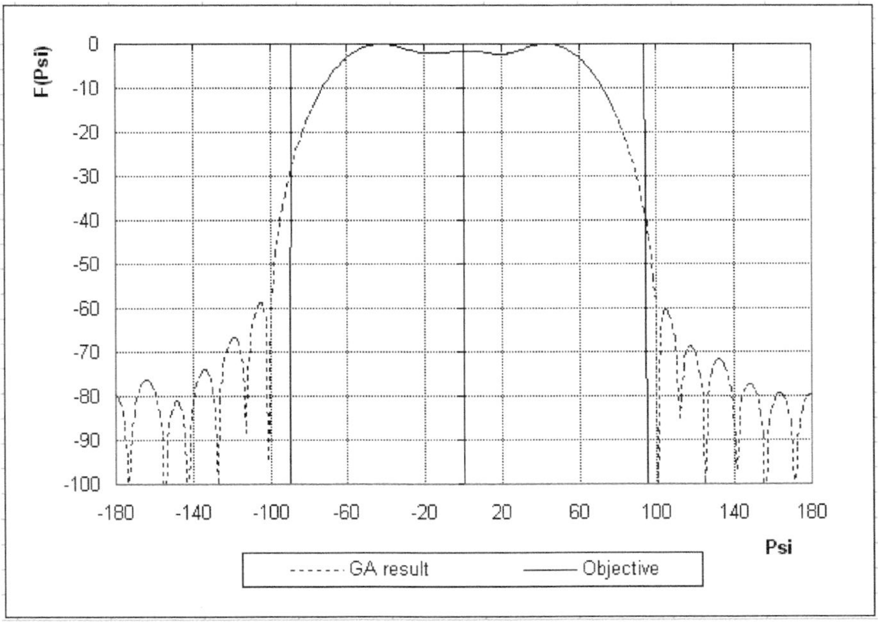

Fig. 5.2 Approximation of the constant array factor with a 15 roots polynomial.

It is noteworthy that the evolutionary method autonomously choose the number of roots used to approximate the objective function: while this is meant to increase the quality of the obtained solution, it also greatly increases the size of the search space, making it more difficult to find an exact solution. One significant advantage of an evolutionary method over the deterministic ones is that the latter ones generate very critical solutions, that is, solutions that cannot be modified, even slightly, without degrading their quality. The evolutionarily generated ones, instead, can undergo greater modifications before losing as much quality as the deterministic ones. This is most probably the effect of these solutions belonging to a population of similar candidate solutions which, during the search process, are selected and mutated: the evolutionary core has a natural tendency to concentrate its population around local maxima which cover large parts of the search space, while very narrow peaks in the fitness function are harder to be detected. The solutions generated with the evolutionary method may undergo further manual optimization. While this is not a desired situation, it may be necessary for some particularly critical problem, anyway comparing this evolutionary approach versus classical methods, it is observable that the latter allows the designer to minimize design time (Figure 5.3).

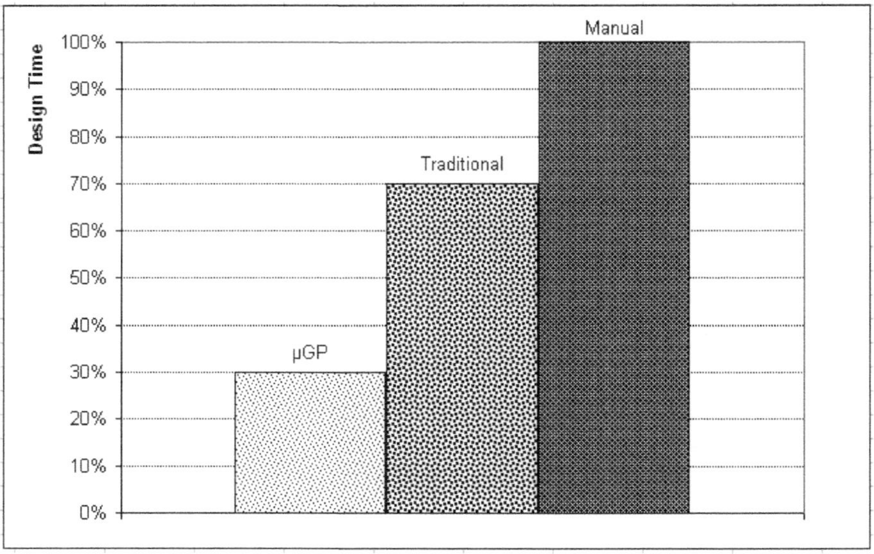

Fig. 5.3 Comparison between different approachs.

5.6 Conclusions

A working environment to perform array antenna synthesis and optimization using an evolutionary approach was presented. A series of experiments were performed, trying to approximate two different objective array factors using different performance measures. The obtained results clearly indicate the need for careful selection of the fitness function within the evolutionary process. They also show that acceptable solutions can be obtained rather quickly and, most importantly, with little human intervention.

The evolutionary tool itself proved very versatile, being able to successfully cope with a problem totally outside of its original application area. This encourages both further investment in the application of evolutionary methods to antenna array syntesis and optimization and development of the tool itself.

Future works will add support for mask specification as well as new fitness measures in the quest for higher-quality solutions. Later on, a graphic interface for simplified usage will be also introduced.

Chapter 6
Drift Correction of Chemical Sensors

Artificial olfaction systems that try to mimic human olfaction by using arrays of gas chemical sensors combined with pattern recognition methods represent a potentially economic tool in many areas of industry such as: perfumery, food and drinks production, clinical diagnosis, health and safety, environmental monitoring and process control. However, successful applications of these systems are still largely limited to specialized laboratories. Among others, sensor drift, the lack of stability over time still limit real industrial setups. This chapter presents and discusses an evolutionary based adaptive drift-correction method designed to work with state-of-the-art classification algorithms. The proposed system exploits a leading-edge evolutionary strategy to iteratively tweak the coefficients of a linear transformation able to transparently transform raw sensors measures in order to mitigate negative effects of the drift. The optimal correction strategy is learned without a-priori models or other hypothesis on the behavior of physical-chemical sensors. Preliminary results have been published in [49].

6.1 Introduction

The human sense of smell is a valuable tool in many areas of industry such as: perfumery, food and drinks production, clinical diagnosis, health and safety, environmental monitoring and process control [65] [156]. Artificial olfaction tries to mimic human olfaction by using arrays of gas chemical sensors combined with pattern recognition (PaRC) methods [121]. When a volatile compound contacts the surface of the sensor array, a set of physical changes modify the electrical properties of each sensor material. Such an electronic disturb can be measured and digitalized to be used as a feature for the specific compound. A preliminary calibration phase is used to train the PaRCalgorithm in order to map each gas concentration or class to the responses from the sensor array. The trained model is then used for identification during later measurements. The classification rate of the PaRCsystem determines the final performance of the electronic olfaction system.

E. Sanchez et al.: Industrial Applications of Evolutionary Algorithms, ISRL 34, pp. 55–74.
springerlink.com

Gas sensor arrays represent a potentially economic and fast alternative to conventional analytical instruments such as gas chromatographs. Considerable research into new technologies is underway, including efforts to use nano-engineering to enhance the performance of traditional resistive Metal Oxide (MOX) sensors. However, successful applications of gas sensor arrays are still largely limited to specialized laboratories [118]. Among others, lack of stability over time and high cost of recalibration still limit the widespread adoption of artificial olfaction systems in real industrial setups [115].

The *gas sensor drift* consists of small and non deterministic temporal variations of the sensor response when it is exposed to the same analytes under identical conditions [115]. This problem is generally attributed to sensors aging [140], but it could be also influenced by a variety of sources including environmental factors or thermo-mechanical degradation and poisoning [79]. As a result, sensors' selectivity and sensitivity decrease, changing the way samples distribute in the data space and thus limiting the ability of operating over long periods. PaRCmodels built in the calibration phase become useless after a period of time, in some cases weeks or a few months. After that time the artificial olfaction system must be completely re-calibrated to ensure valid predictions [5]. Up to now, it is impossible to fabricate chemical sensors without drift. In fact, drift phenomena afflict almost all kinds of sensors [123][31][113]. Sensor drift should be therefore taken into account and compensated in order to achieve reliable measurement data from the sensor array.

Algorithms to mitigate negative effects of the gas sensor drift are not new in the field, with the first attempt to tackle this problem dated back to the early 90s [121, chap. 13]. Notwithstanding, the study of the sensor drift is still a challenging task for the chemical sensor community [121][115]. Solutions proposed in the literature can be grouped in three main categories: (i) periodic calibration, (ii) attuning methods, and (iii) adaptive models.

Retraining the PaRCmodel by using a single calibrant or a set of calibrants is perhaps the only robust method to mitigate drift effects even in presence of sensor drift over an extremely long period of time [142]. However, calibration is the most time-intensive method for drift correction since it requires system retraining and additional costs. Hence, it should be used sparingly. Moreover, while this approach is rather simple to implement for physical sensors where the quantity to be measured is exactly known, chemical sensors pose a series of challenging problems. Indeed, in chemical sensing, the choice of the calibrant strongly depends on the specific application especially when the sensing device is composed of a considerable number of cross-correlated sensors [74][73]. This leads to loss of generalization and lack of standardization which, on the contrary, would be highly required by industrial systems.

Attuning methods try to separate and reject drift components from real responses. They can provide significant improvements in the classification rate over a fixed time period, and may also allow to obtain real responses to be used in gas quantitative analysis. Attempts to attune the PaRCmodel by performing component correction based on Principal Component Analysis (PCA) [7][153], Independent Components Analysis (ICA) [110], Canonical Correlation Analysis (CCA), or

Partial Least Squares (PLS) [68] have received considerable attention in the sensor community. Orthogonal Signal Correction (OSC) was recently demonstrated to be one of the best methods to attune PaRCmodels and to compensate drift effects [115]. However, such techniques do not completely solve the problem. One of the main drawbacks is the need of a set of calibration data containing a significant amount of drift allowing to precisely identify the set of components to be corrected or rejected. This might be not the case in industrial setups where calibration data are collected over a short period of time. Moreover, adding new analytes to the recognition library represents a major problem since rejected components might be necessary to robustly identify these new classes. Finally, these methods contain no provisions for updating the model and thus may ultimately be invalidated by time evolving drift effects.

Adaptive models try to adapt the PaRCmodel by taking into account pattern changes due to drift effects. Neural networks such as self-organizing maps (SOMs) [103][166] or adaptive resonance theory (ART) networks [157][99] have been frequently used in the past. They have the advantage of simplicity because no recalibration is required. Yet, two main weaknesses can be identified. First, a discontinuity in response between two consecutive exposures (regardless of the time interval between the exposures) would immediately invalidate the PaRCmodel and would prevent adaptation. Second, a key to obtain reliable results is to set appropriate thresholds for choosing the winning neuron, and this typically requires a high number of training samples owing the complexity of the network topology. Moreover, they are limited to gas classification applications. Whenever both classification and gas quantitative analysis is required, current adaptive methods can be hardly applied to obtain reliable gas concentration measurements [78].

In this chapter we present and discuss an evolutionary based adaptive unsupervised drift-correction methodology designed to work with state-of-the-art classification algorithms. The term unsupervised refers to the fact that drift correction is obtained without considering any specific drift model. Drift effects are directly learned from the set of unlabeled raw measures obtained from the sensor array. This work improves our previous attempt to apply evolutionary methods in the drift correction process [49]. A linear transformation is applied to raw sensor's features to compensate drift effects. Such linear transformation is continuously and slowly evolved to follow drift effects. Evolution is achieved through a covariance matrix adaptation evolutionary strategy (CMA-ES), perfectly suited for solving difficult optimization problems in continuous domain. Compared to existing adaptive solutions, the proposed approach is able to transparently adapt to changes in the sensors' responses even when the number of available samples is not high and new classes of elements are introduced in the classification process at different time frames. Experimental results demonstrate that the suitability of the proposed methodology does not depend on the exploited classifier.

6.2 Method and Theory

he basic steps and concepts of the proposed drift correction process are summarized in Figure 6.1.

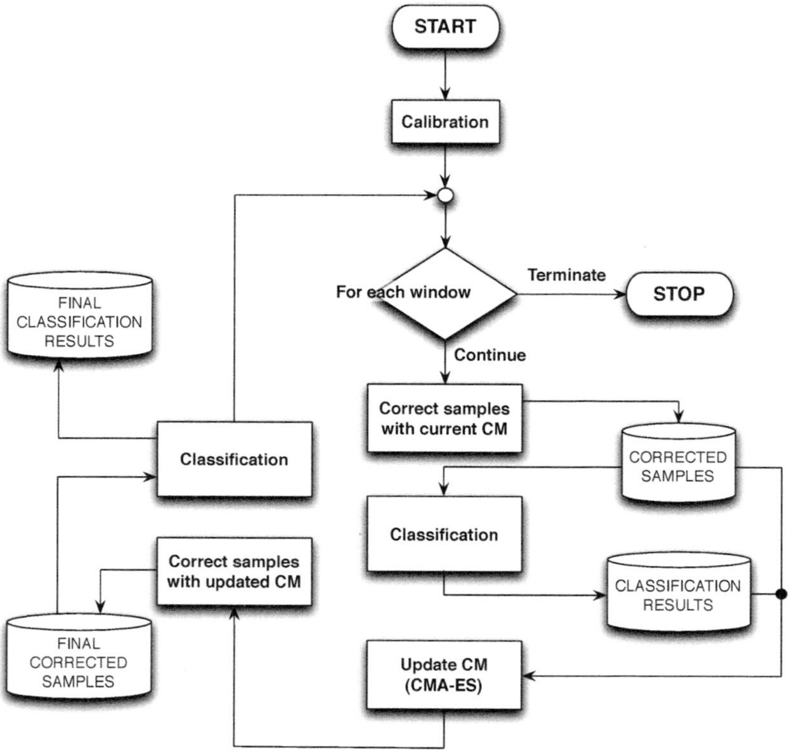

Fig. 6.1 Conceptual steps of the drift correction process

As common in artificial olfaction systems a preliminary calibration phase is used to collect a set of training samples for m different classes y_i ($i \in [1,m]$), each one identifying a specific gas compound. Training samples are used to train a classifier able to map a generic sample $\mathbf{x} \in \mathbb{R}^n$ (where n is the number of sensors in the array) into one of the m available classes:

$$\mathscr{C} : \mathbf{x} \rightarrow \{y_1, y_2, \ldots, y_m\} \tag{6.1}$$

Any type of classification algorithm can be theoretically plugged into this system. The idea behind the proposed drift correction method is to reduce variations in the sensors response caused by the sensor drift, thus augmenting the validity window of the classification model.

Once the calibration phase is concluded the system is ready to accept samples to be analyzed and classified. The analysis is performed considering windows of samples. A *window* (W) is a collection of k consecutive measurements obtained by the same sensor array, where the drift may be assumed linear. Windows are not necessarily associated to measurement sessions: a single measurement session may be split into multiple windows; and multiple sessions may be grouped into a single window depending on the specific application and measurement setup. For example, in a laboratory where the same expensive equipment is shared between different research groups, consecutive sessions of measurements could be grouped in the same window, while sessions for the same project that take place after the equipment has been used for another research could be put in a separate window. We denote with $\mathbf{x}_{i,j} \in \mathbb{R}^n$ the jth sample of the ith window W_i. Within a window samples are ordered with ascending sampling time and the same happens for different windows.

According to the definition of Section 6.1 we assume that the sensor drift causes changes in the sensors' response slowly over the time and that both its direction and intensity for each considered sample are not randomly distributed.

For each window W_i the drift correction process performs five computational steps:

1. Each sample $\mathbf{x}_{i,j} \in W_i$ is corrected by applying a *correction factor* (cf) able to mitigate the drift effect (see Section 6.2.1). The result is a set of *corrected samples* denoted as: $\mathbf{xc}_{i,j} \in \mathbb{R}^n$);
2. Each corrected sample $\mathbf{xc}_{i,j}$ is classified using the classifier $\mathscr{C}$ of equation 6.1 trained during the calibration phase (see Section 6.2.2);
3. Corrected samples and classification results are used in an evolutionary process to adapt the current correction factor to the changes of the sensor drift observed in the current window (see Section 6.2.3);
4. Each sample $\mathbf{x}_{i,j} \in W_i$ is corrected again by applying the updated correction factor computed during step 4;
5. Corrected samples are classified again, and the final classification results are provided as outcome of the system.

The following subsections provide details on how the different steps are implemented.

6.2.1 Correction Factor

By considering a sample $\mathbf{x}_{i,j} \in \mathbb{R}^n$ as a point in the n-dimensional space of the sensor array features, the drift effect represents a translation of the point along a preferred direction. Under the hypotheses that in the very short term the variation imposed by the drift is small we can approximate it with a linear translation [7] and we can therefore envision to correct it by applying a linear transformation.

Given a sample $\mathbf{x}_{i,j} \in W_i$ the corrected sample $\mathbf{xc}_{i,j}$ is therefore computed as :

$$\mathbf{xc}_{i,j} = \mathbf{x}_{i,j} + \underbrace{\mathbf{x}_{i,j} \times \mathbf{M}_i}_{\text{correction factor}} \tag{6.2}$$

where $\mathbf{M}_i \in \mathbb{R}^{n \times n}$ is the *correction matrix* for the window W_i generating a correction factor for each feature of the sample obtained as a linear combination of the values of all features in the sample. Considering all features when computing the correction factor allows us to take into account correlations among sensors in the drift phenomena.

The correction factor of feature i of a sample $\mathbf{x}$ can be therefore computed as:

$$cf_i = x[1] \cdot M[1][i] + \ldots + x[n] \cdot M[n][i] \tag{6.3}$$

The correction matrix for the first window ($\mathbf{M}_1$) is initially set to the null matrix, i.e., no correction is applied immediately after calibration.

6.2.2 Classification

Once the drift has been compensated, corrected samples can be classified. State-of-the-art classifiers (e.g., k-NN , Random Forests , etc. [51]) can be applied in this phase without need of modifications to the standard implementations . The possibility of working with any type of external classifier represents one of the strengths of the proposed method, allowing to choose the best PaRCmodel based on the specific application.

6.2.3 Correction Factor Optimization

The correction matrix $\mathbf{M}_i$, used to correct samples of a window W_i, is continuously adapted when passing from a window to the next one. The overall goal of this optimization process is to update $\mathbf{M}_i$ on the basis of the information provided by the samples of W_i in order to follow the evolution of the drift and therefore be prepared for the analysis of the next window W_{i+1}.

The adaptation is obtained using the CMA-ES, a stochastic population-based search method in continuous search spaces, aiming at minimizing an objective function $f : \mathscr{S} \subseteq \mathbb{R}^p \rightarrow \mathbb{R}$ in a black-box scenario (see 6.4 for specific details).

In our specific application the solution computed by the CMA-ES during the elaboration of the window W_i identifies the candidate correction matrix for the window W_{i+1} ($\mathbf{M}_{i+1}$). We denote with $\mathbf{M}^{\mathbf{s}}$ the correction matrix obtained from the solution $\mathbf{s} \in \mathscr{S} \subseteq \mathbb{R}^{p=n \cdot n}$ by computing each element as follows:

$$M^{\mathbf{s}}[i][j] = s[(i-1) \cdot n + j], (i \in [1,n], j \in [1,n]) \tag{6.4}$$

The objective function applied in the optimization process, computed considering the set of samples belonging to window W_i, is expressed as:

$$f_i(\mathbf{s}) = \sum_{j=0}^{|W_i|-1} D\left(\mathbf{x}_{i,j} + \mathbf{M}^{\mathbf{s}} \times \mathbf{x}_{i,j}, \mu_{\mathscr{C}(\mathbf{x_{i,j}})}\right) \tag{6.5}$$

It computes the sum of the distances (D) of each corrected sample in W_i ($\mathbf{x}_{i,j} + \mathbf{M}^{\mathbf{s}} \times \mathbf{x}_{i,j}$) from the centroid of the related class in the training set ($\mu_{\mathscr{C}(\mathbf{x_{i,j}})}$). The centroid of a class y is computed as follows:

$$\mu_y = \frac{\sum_{i=1}^{|y|} \mathbf{t}_i^y}{|y|} \tag{6.6}$$

where $|y|$ is the number of training samples for the class y and $\mathbf{t}_i^y$ is the ith training sample for the class. The function $f_i(\mathbf{s})$ tries to measures how corrected samples tend to deviate from the class distributions learnt by the classifier during the calibration phase.

The evolutionary process stops the optimization based on the following stop conditions:

1. The optimum value of the objective function has been reached. Depending on the type of distance function D considered in equation 6.5 (see Section 6.2.4), the optimal value of the objective function can be set to zero or to a lower bound indicating that all corrected samples have been collapsed into a region closed to the the centroid of the class they belong to. Due to the complexity of the optimization process this condition cannot be always reached;
2. During the optimization all candidate solutions in the current population P_c have a value of the objective function differing from that of the other candidates less than a predefined threshold ω_{min}:

$$f_i(\mathbf{s_x}) - f_i(\mathbf{s_y}) < \omega_{min} \qquad \forall x, y \in P_c \tag{6.7}$$

3. The step size σ_{cur} of the CMA-ES (see 6.4) increases more than a predefined threshold $\bar{\sigma}_{max}$ with respect to its initial value σ_{ini}, i.e., the optimization process is trying to explore an area in the search space that is too large; or σ_{cur} decreases more than a predefined threshold $\bar{\sigma}_{min}$, i.e., the optimization process is trying to explore a local minima:

$$\begin{cases} |\sigma_{ini} - \sigma_{cur}| > \bar{\sigma}_{max} \\ |\sigma_{ini} - \sigma_{cur}| < \bar{\sigma}_{min} \end{cases} \tag{6.8}$$

The initial step size σ_{ini} is used to sample the search space around an initial search point (i.e., a randomly chosen value or a previous solution).

Together with the three defined stop conditions, the optimization is also interrupted if a maximum number of generations has been reached.

6.2.4 Distance Functions

Four types of distances have been used in this work to compute the objective function of equation 6.5:

- *Mahalanobis distance*: the Mahalanobis distance computes the distance between two samples by taking into account how samples distribute in the space. It allows to overcome problems deriving by non spherical distributions of samples:

$$D_m(\mathbf{x}, \mu_c) = \sqrt{(\mathbf{x} - \mu_c) \cdot \mathbf{Cov}^{-1} \cdot (\mathbf{x} - \mu_c)^T} \tag{6.9}$$

  where $\mathbf{Cov}^{-1}$ is the inverse of the covariance matrix for the samples of the training set of class c.
- *Exponential distance*: the Mahalanobis distance of the sample $\mathbf{x}$ is exponentially scaled, as follows:

$$D_x(\mathbf{x}, \mu_c) = e^{D_m(\mathbf{x}, \mu_c)} \tag{6.10}$$

  It exponentially penalizes samples that are moved far from the related centroid.
- *Linear step distance*: the distance of the sample $\mathbf{x}$ from the centroid of its class is computed as a step function as follows:

$$D_{ls}(\mathbf{x}, \mu_c) = \begin{cases} 0 & 0 \leq D_m(\mathbf{x}, \mu_c) \leq D^c_{m_{max}} \\ \frac{D_m(\mathbf{x}, \mu_c)}{D^c_{m_{max}})} - 1 & D^c_{m_{max}} < D_m(\mathbf{x}, \mu_c) \leq 2D^c_{m_{max}} \\ 10^3 & 2D^c_{m_{max}} < D_m(\mathbf{x}, \mu_c) \end{cases} \tag{6.11}$$

  where $D^c_{m_{max}}$ is the maximum Mahalanobis distance of samples of the training set of the class c from the related centroid. This step function gives maximum importance to samples close to the centroid of the related class ($D_{ls}(\mathbf{x}, \mu_c) = 0$) while strongly penalizes samples that are moved far from the centroid of the related class ($D_{ls}(\mathbf{x}, \mu_c) = 10^3$). In the region between the two cases the distance is increased linearly.
- *Exponential step distance*: similarly to D_{ls} the distance is computed as a step function as follows:

$$D_{xs}(\mathbf{x}, \mu_c) = \begin{cases} 0 & 0 \leq D_m(\mathbf{x}, \mu_c) \leq D^c_{m_{max}} \\ e^{Dm(\mathbf{x}, \mu_c)} & D^c_{m_{max}} < D_m(\mathbf{x}, \mu_c) \leq 2D^c_{m_{max}} \\ e^{2Dm^c_{max}} & 2D^c_{m_{max}} < D_m(\mathbf{x}, \mu_c \end{cases} \tag{6.12}$$

  the main difference w.r.t. D_{ls} is the way samples far from the centroid are penalized.

The choice of the best distance function to use depends on the considered dataset. This represents a degree of freedom that allows to tune the drift correction system for the specific application.

6.3 Case Studies and Experimental Results

The proposed methodology has been validated on a set of experiments performed on two datasets: the first composed of simulated data artificially generated, while the second composed of samples obtained from a real application.

The full correction system has been implemented as a combination of Perl and C code. A pool of four classifiers have been considered: k-Nearest Neighbors (*k*NN), Partial Least Square Discriminant Analysis (PLS), Neural Networks (NNET) and Random Forest (RF). All classifiers have been implemented using the Classification And REgression Training (CARET) package of *R*, a free and multi-platform programming language and software environment widely used for statistical software development and data analysis. Details on the specific implementation of the classifiers are available in [92]. The performance of the prediction model of each classifier has been tuned and optimized by performing leave-group-out-cross-validation (LGOCV). For each classifier 50 folds of the training set have been generated with 95% of samples used to train the model while the remaining ones used as test data. The size of the grid used to search the tuning parameters space for each classifier (e.g., k for *K*NN) has been set to 5. This represents a good compromise in terms of computational time of the training phase and optimization results.

Table 6.1 Parameters resulting from the tuning of each classifier

Optimal classifiers parameters for artificial data set			
Classifier	**Parameter**	**Description**	**Value**
*k*NN	*k*	Number of nearest neighbors	37
PLS	ncomp	Number of components one wishes to fit	4
NNET	size	Number of units in the hidden layer	3
	decay	Parameter of weight decay	0.1
RF	mtry	Number of variables randomly sampled as candidates at each split	2
Optimal classifiers parameters for real data set			
Classifier	**Parameter**	**Description**	**Value**
*k*NN	*k*	Number of nearest neighbors	21
PLS	ncomp	Number of components one wishes to fit	6
NNET	size	Number of units in the hidden layer	5
	decay	Parameter of weight decay	0.03
RF	mtry	Number of variables randomly sampled as candidates at each split	4

The optimal parameters obtained from the classifiers tuning phase are reported in table 6.1.

6.3.1 Artificial Dataset

For a preliminary evaluation study we tested the proposed drift correction methodology on simulated data composed of a given number of independent, uncorrelated and randomly distributed Gaussian clusters. The Gaussian model is often regarded as a benchmark in literature for gas chemical sensors data analysis [54]. It therefore provides an effective platform for testing the validity of the proposed approach. Simulated data allow to control the parameters influencing the drift correction capability such as, feature space dimensionality n, number of classes m, separation among clusters α (given in standard deviation units) and drift direction/intensity.

6.3.1.1 Experimental Setup

We considered a data set of 1000 samples belonging to 5 different classes ($m = 5$). Each sample includes 6 features ($n = 6$) simulating a sensor array composed of 6 sensors. The centroid of each class c is randomly drawn according to a multivariate normal distribution in n dimensions $\mu_c = \mathcal{N}\left(\mathbf{0}, \frac{\alpha^2}{2n}\mathbf{I}\right)$ ($\alpha = 12$ in our specific case). Using the term $\frac{\alpha^2}{2n}$ as scaling factor of the variance, the expectation value of the square distance between any two centroids is equal to α^2 independently of n. This allows to have enough separation among classes to build efficient classifiers. In order to control the minimum clusters separation we discarded simulations where, due to the randomness of the process, any two centers are closer than $\alpha/2$. For each class, we generated 250 Gaussian distributed samples with unit variance affected by a drift linear in time according to the following equation:

$$\mathbf{x}(c,t) = \mathcal{N}(\mu_c, \mathbf{I}) + \underbrace{\left(\frac{t}{h} \cdot \mathbf{u_d}\right)}_{\text{drift effect}} \tag{6.13}$$

where h represents a scaling factor for the discrete time t (h has been set to 40 in our specific case to guarantee a significant amount of drift). The term $\mathbf{u_d}$ represents a randomly generated unitary vector in the n-dimensional space describing the direction of the drift applied to each sample of the dataset. In our simulated data all classes are linearly drifted in the same direction, and samples of the different classes are uniformly distributed in time to present similar drift conditions. The effect of the drift is evident by looking at the projection over the first two principal components of the PCA reported in Figure 6.2.

The experimental session included 100 runs of the drift-correction process for each of the four objective functions based on the distances introduced in Section 6.2.4. The first 100 samples of the data set have been used as training data for the PaRCmodel, while the remaining 900 samples have been used as test set to be analyzed. The test set has been processed splitting the data in windows of 50 samples.

6.3.1.2 Results and Discussion

Table 6.2 shows the performance of the proposed system for the five considered classifiers and the four considered objective functions. Results are provided in terms of classification rate on each of the 18 windows and total classification rate (T. Cr.). To better highlight the benefits of the correction process, Table 6.2 reports both the classification rate of each classifier when no correction is applied and the one

Table 6.2 Performance of the drift correction system in terms of classification rate on the artificial data set

Classifier		Classification rate over windows W_i																		T.Cr
		W_1	W_2	W_3	W_4	W_5	W_6	W_7	W_8	W_9	W_{10}	W_{11}	W_{12}	W_{13}	W_{14}	W_{15}	W_{16}	W_{17}	W_{18}	
Classifiers without drift correction																				
*k*NN		1.00	0.96	0.98	0.88	0.88	0.80	0.90	0.78	0.78	0.68	0.66	0.62	0.68	0.60	0.60	0.62	0.60	0.58	0.75
NNET		1.00	1.00	1.00	0.96	1.00	0.98	1.00	0.98	0.98	0.94	0.94	0.82	0.92	0.90	0.86	0.86	0.86	0.82	0.93
PLS		1.00	0.96	0.98	0.88	0.92	0.82	0.86	0.80	0.64	0.64	0.56	0.52	0.38	0.32	0.26	0.30	0.26	0.24	0.63
RF		0.86	0.88	0.86	0.80	0.80	0.80	0.78	0.76	0.76	0.74	0.64	0.62	0.62	0.56	0.48	0.48	0.44	0.46	0.68
Drift correction using the Mahalanobis distance D_m																				**ΔT.Cr**
*k*NN	Avg	1.00	1.00	0.96	0.99	0.96	0.97	0.95	0.95	0.97	0.95	0.94	0.91	0.89	0.86	0.84	0.81	0.79	0.78	+0.17
	C.I.	.006	.015	.012	.015	.016	.020	.019	.021	.024	.026	.026	.027	.027	.028	.030	.029	.029	.030	
NNET	Avg	0.98	1.00	0.99	0.96	0.97	0.99	0.98	0.99	0.99	0.95	0.97	0.92	0.94	0.91	0.91	0.89	0.84	0.85	+0.02
	C.I.	.002	.000	.003	.003	.003	.002	.004	.002	.002	.013	.012	.015	.015	.015	.017	.020	.026	.029	
PLS	**Avg**	**1.00**	**1.00**	**1.00**	**0.96**	**0.99**	**0.97**	**0.98**	**0.98**	**0.99**	**0.98**	**0.99**	**0.96**	**0.94**	**0.94**	**0.90**	**0.84**	**0.80**	**0.80**	**+0.31**
	C.I.	**.001**	**.000**	**.001**	**.005**	**.004**	**.005**	**.006**	**.005**	**.002**	**.004**	**.006**	**.013**	**.015**	**.017**	**.020**	**.025**	**.030**	**.031**	
RF	Avg	0.99	0.97	1.00	0.99	0.91	0.88	0.84	0.93	0.87	0.85	0.84	0.81	0.83	0.82	0.81	0.82	0.80	0.79	+0.19
	C.I.	.002	.002	.001	.002	.007	.006	.005	.011	.008	.007	.006	.005	.006	.006	.009	.006	.010	.011	
Drift correction using the linear step distance D_{ls}																				**ΔT.Cr**
*k*NN	Avg	0.98	0.92	0.94	0.88	0.92	0.89	0.88	0.86	0.88	0.85	0.83	0.81	0.79	0.78	0.76	0.76	0.73	0.73	+0.09
	C.I.	.006	.015	.012	.015	.016	.020	.019	.021	.024	.026	.026	.027	.027	.028	.030	.029	.029	.030	
NNET	Avg	0.97	0.92	0.94	0.88	0.89	0.87	0.87	0.85	0.86	0.82	0.82	0.78	0.79	0.77	0.76	0.76	0.73	0.71	-0.10
	C.I.	.005	.013	.011	.015	.019	.022	.021	.024	.025	.025	.029	.028	.031	.029	.031	.032	.034	.036	
PLS	Avg	0.98	0.91	0.93	0.88	0.91	0.87	0.84	0.84	0.82	0.78	0.76	0.74	0.70	0.67	0.66	0.66	0.62	0.61	+0.16
	C.I.	.006	.016	.013	.016	.020	.023	.025	.026	.032	.034	.037	.039	.040	.041	.043	.043	.044	.045	
RF	Avg	0.98	0.92	0.94	0.92	0.87	0.85	0.81	0.81	0.79	0.78	0.77	0.75	0.74	0.73	0.72	0.72	0.69	0.68	+0.12
	C.I.	.006	.008	.009	.012	.015	.016	.014	.018	.016	.017	.017	.017	.020	.019	.021	.024	.026	.026	
Drift correction using the exponential distance D_x																				**ΔT.Cr**
*k*NN	Avg	1.00	0.99	0.98	0.92	0.95	0.92	0.90	0.88	0.88	0.84	0.83	0.80	0.77	0.77	0.76	0.75	0.70	0.69	+0.10
	C.I.	.000	.002	.003	.015	.015	.016	.018	.022	.024	.028	.028	.031	.032	.032	.033	.033	.035	.035	
NNET	Avg	1.00	1.00	1.00	0.94	0.96	0.92	0.80	0.75	0.77	0.72	0.70	0.66	0.66	0.64	0.64	0.62	0.60	0.59	-0.16
	C.I.	.000	.002	.002	.007	.014	.020	.034	.040	.038	.039	.039	.037	.038	.035	.037	.035	.035	.037	
PLS	Avg	1.00	1.00	0.99	0.93	0.95	0.89	0.83	0.79	0.77	0.74	0.72	0.67	0.65	0.66	0.63	0.60	0.57	0.55	+0.14
	C.I.	.000	.000	.003	.013	.014	.024	.031	.034	.037	.038	.037	.040	.038	.037	.040	.041	.040	.038	
RF	Avg	0.92	0.90	0.91	0.90	0.84	0.82	0.80	0.83	0.78	0.76	0.74	0.72	0.70	0.69	0.66	0.67	0.64	0.62	+0.09
	C.I.	.006	.008	.011	.016	.015	.016	.013	.021	.020	.024	.022	.028	.028	.029	.029	.030	.031	.032	
Drift correction using the exponential step distance D_{xs}																				**ΔT.Cr**
*k*NN	Avg	1.00	0.99	0.98	0.92	0.92	0.90	0.91	0.88	0.89	0.87	0.86	0.81	0.80	0.79	0.78	0.76	0.73	0.74	+0.11
	C.I.	.000	.003	.006	.011	.016	.016	.016	.019	.021	.023	.024	.027	.025	.028	.028	.026	.027	.028	
NNET	Avg	0.98	0.98	0.98	0.93	0.94	0.94	0.94	0.93	0.92	0.89	0.88	0.82	0.82	0.81	0.80	0.80	0.77	0.76	-0.05
	C.I.	.001	.004	.005	.007	.014	.016	.015	.016	.016	.019	.025	.026	.028	.030	.030	.032	.033	.037	
PLS	Avg	0.99	0.99	0.99	0.94	0.93	0.91	0.91	0.90	0.88	0.87	0.85	0.82	0.79	0.75	0.73	0.72	0.69	0.67	+0.22
	C.I.	.002	.003	.003	.013	.017	.020	.019	.023	.026	.026	.028	.032	.032	.034	.037	.036	.038	.041	
RF	Avg	0.99	0.96	0.99	1.00	0.95	0.95	0.92	0.95	0.91	0.89	0.87	0.83	0.83	0.83	0.80	0.81	0.79	0.76	+0.21
	C.I.	.002	.002	.003	.002	.008	.011	.013	.012	.017	.017	.015	.016	.015	.019	.019	.019	.017	.019	

considering the correction system. Results for the correction system are produced in terms of average classification rate over the 100 considered runs (Avg). In order to evaluate the stability of the results over the different runs, for each average value is reported the related confidence interval (C.I.), computed considering 95% level of confidence. The total classification rate is expressed in this case as the variation w.r.t. the one of the classifier without correction.

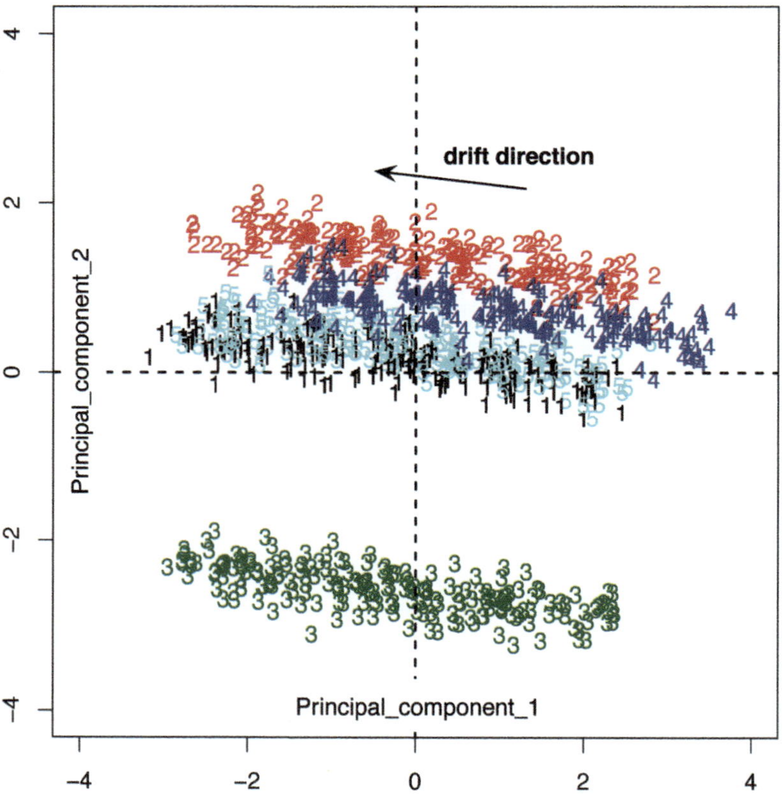

Fig. 6.2 Projection of the first two principal components of the PCA computed for the artificially generated dataset.

Results provided in Table 6.2 confirm that in general, for all considered classifiers, the drift correction process allows to improve the classification rate with results that are quite stable over the different runs. In particular, the two objective functions based on the Mahalanobis distance (D_m) and the exponential step distance (D_{xs}) seem to provide better results. Among the different classifiers, NNET gained lower improvement due to the fact that the classification rate was already quite high even without applying any correction. On the contrary, we observed the most significant improvement w.r.t. the classifier, when applied on raw measures, with PLS system corrected with the objective function based on the Mahalanobis distance.

Figure 6.3 graphically compares the performance of the proposed drift correction method with the Orthogonal Signal Correction (OSC) that, as introduced in Section 6.1, represents a state-of-the-art attuning method to perform drift correction. OSC has been implemented using the osccalc.m function of the PLS toolbox package (ver. 5.5) for MATLAB environment (64 bit, ver. 7.9). For the experiments we chose to remove one orthogonal component. Results are evaluated considering the PLS classifier corrected with the objective function based on the Mahalanobis distance. Since the size of the training set strongly influences the effectiveness of this approach, we provided results considering different values for the training set size (100 samples for osc-100 and 200 samples for osc-200) [115]. The proposed results clearly show how the proposed method outperforms the OSC requiring a reduced set of training data.

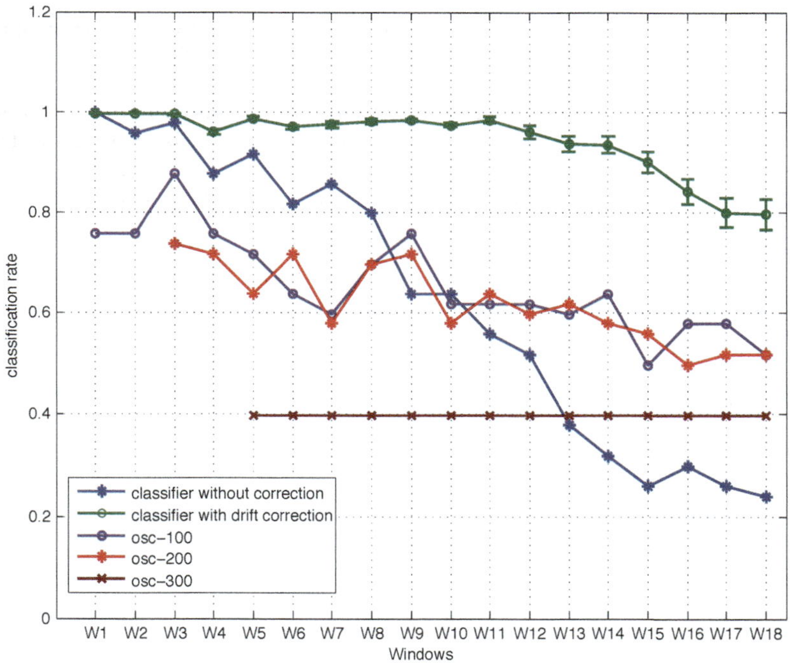

Fig. 6.3 Comparison of the proposed drift correction systems with the OSC for the PLS classifier with the objective function using the Mahalanobis distance D_m.

Finally, to show the ability of the correction process to actually remove the drift component from the considered samples, Figure 6.4 graphically shows the projection over the first two principal components for the corrected dataset for one of the runs performed with the PLS classifier using the Mahalanobis distance(Figure 6.4-a) and for the original data without drift (Figure 6.4-b). The last set of data was stored during the generation of the artificial dataset before inserting the drift component (see equation 6.13). Both plots have been generated using the same projection to allow comparison. The figure confirms how the drift observed in Figure 6.2 has been strongly mitigated allowing a distribution of samples that approximate the one without drift. This is an important results allowing to perform quantitative gas analysis and further examinations on the corrected data overcoming one of the main problems of previous adaptive correction methods (see Section 6.1).

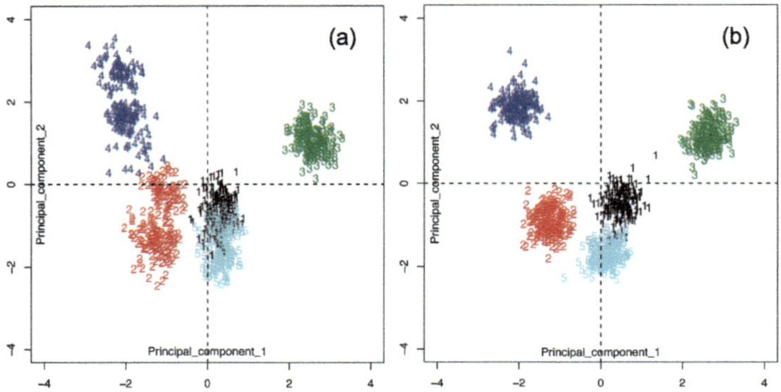

Fig. 6.4 Comparison of the corrected data set (a) with the original data without drift for the artificial data set (b), using PLS classifier

6.3.2 *Real Dataset*

To additionally validate the proposed approach we also performed a set of experiments on a real data set collected at the *SENSOR Lab*, an Italian research laboratory specialized in the development of chemical sensor arrays[1]. All data have been collected using an EOS835 *electronic nose* composed of 6 chemical MOX sensors: further information on sensors and equipment used can be found in [118] and its references. The goal of the experiment is to determine whether the EOS835 can identify five pure organic vapors: ethanol (class 1), water (class 2), acetaldehyde (class 3), acetone (class 4), ethyl acetate (class 5). All these are typical chemical compounds to be detected in real-world applicative scenarios.

6.3.2.1 Experimental Setup

A total of five different sessions of measurements were performed over one month to collect a dataset of 545 samples, a high value compared to other real datasets reported in the literature. While the period of time was not very long, it was enough to obtain data affected by a certain amount of drift. Not all classes of compounds have been introduced since the first session mimicking a common practice in real-world experiments: samples of classes 1 and 2 have been introduced since the beginning;

[1] `http://sensor.ing.unibs.it/`

class 3 is first introduced during the second session, one week later; first occurrences of classes 4 and 5 appear only during the third session, 10 days after the beginning of the experiment. Classes are not perfectly balanced in terms of number of samples, with a clear predominance of classes 1, 2 and 3 over classes 4 and 5. All these peculiarities make this dataset complex to analyze allowing us to stress the capability of the proposed correction system. The effect of the drift is evident by looking at the projection over the first two principal components of the PCA reported in Figure 6.5.

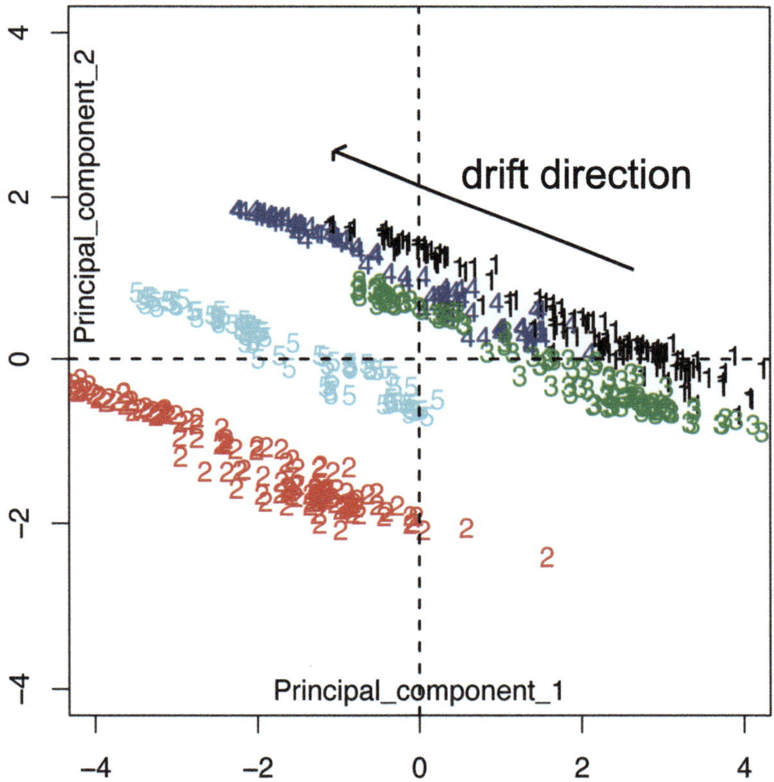

Fig. 6.5 Projection of the first two principal components of the PCA computed for the real dataset.

As for the artificial dataset the experimental session included 100 runs of the drift-correction process for each of the four considered objective functions. The first 20 samples of each class have been used as training data for the PaRCmodel, while the remaining 445 samples have been used as test set. The drift correction process has been applied to windows of 100 samples, with the last one of 45 samples. The bigger size of the windows compared to the artificial dataset is required to tackle the additional complexity of the real data.

6.3.2.2 Results

Table 6.3 summarizes the performance of the drift correction system on the real data set.

Results immediately highlight how the correction process for this particular experiment is harder than that for the artificial data. Main difficulties are connected to the fact that samples from different classes are introduced non homogeneously over the time and the initial interclass distance among the centroids is not enough to avoid partial overlapping of the classes. Moreover, the increased size of the windows increases the effort required by the CMA-ES to compute the appropriate correction matrices. However, the exponential step distance steal produces interesting improvements in the classification rate. Looking also at the results of the artificial dataset this distance seems the best compromise to work with generic data.

PLS corrected with the objective function based on the exponential step distance is the classifier that gained better improvements. Figure 6.6 compares again the results for this case with the correction obtained applying the OSC. This time due to the limited amount of samples, a single case with 100 samples of training has been considered. Again the proposed drift correction approach performs better than the OSC.

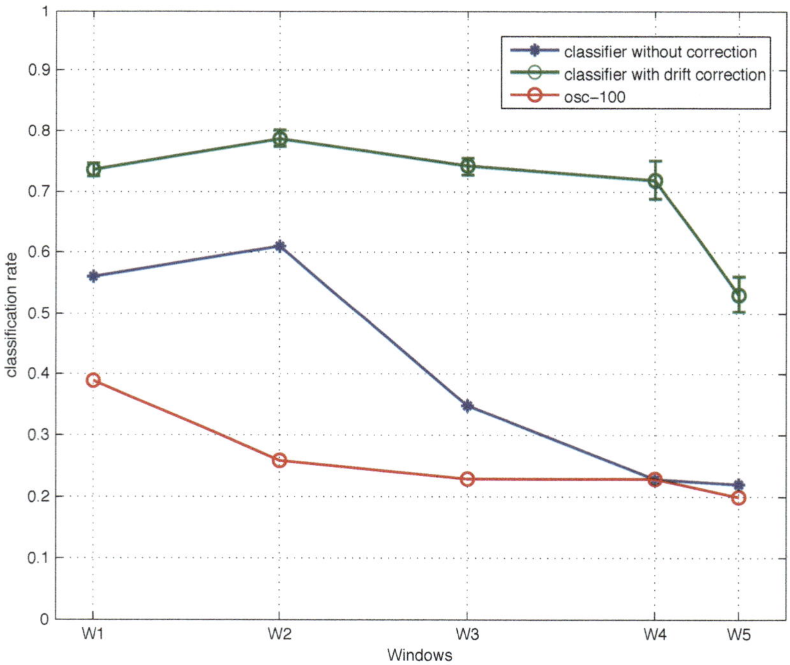

Fig. 6.6 Comparison of the proposed drift correction systems with the OSC for the PLS classifier with objective function using the exponential step distance D_{xs}

[!h]

Table 6.3 Performance of the drift correction system in terms of classification rate on the artificial real set

Classifier		Classification rate over windows W_i: W_1	W_2	W_3	W_4	W_5	T.Cr
Classifiers without drift correction Dm							
*k*NN		0.63	0.54	0.35	0.32	0.31	0.45
NNET		0.56	0.65	0.63	0.47	0.36	0.55
PLS		0.56	0.61	0.35	0.23	0.22	0.42
RF		0.86	0.86	0.82	0.70	0.69	0.80
Drift correction using the Mahalanobis distance D_m							**ΔT.Cr**
*k*NN	Avg	0.66	0.62	0.53	0.55	0.52	+0.13
	C.I.	.004	.025	.015	.027	0.32	
NNET	Avg	0.28	0.54	0.52	0.50	0.49	-0.09
	C.I.	.009	.009	.016	.022	.026	
PLS	Avg	0.40	0.41	0.28	0.30	0.37	-0.07
	C.I.	.016	.015	.018	.021	.024	
RF	Avg	0.90	0.78	0.80	0.80	0.80	+0.02
	C.I.	.003	.003	.002	.000	.000	
Drift correction using the linear step distance D_{ls}							**ΔT.Cr**
*k*NN	Avg	0.89	0.72	0.55	0.56	0.49	+0.21
	C.I.	.006	.022	.023	.033	.033	
NNET	Avg	0.81	0.70	0.71	0.68	0.54	+0.16
	C.I.	.013	.023	.030	.046	.038	
PLS	Avg	0.86	0.67	0.53	0.55	0.41	+0.21
	C.I.	.007	.017	.025	.031	.029	
RF	Avg	0.95	0.91	0.86	0.85	0.83	+0.08
	C.I.	.002	.012	.010	.023	.031	

Classifier		Classification rate over windows W_i: W_1	W_2	W_3	W_4	W_5	T.Cr
Drift correction using the exponential distance D_x							**ΔT.Cr**
*k*NN	Avg	0.20	0.24	0.28	0.22	0.27	-0.21
	C.I.	.000	.009	.021	.017	.020	
NNET	Avg	0.02	0.41	0.26	0.31	0.28	-0.30
	C.I.	.010	.022	.018	.029	.022	
PLS	Avg	0.20	0.26	0.29	0.26	0.30	-0.16
	C.I.	.000	.015	.021	.020	.022	
RF	Avg	0.60	0.64	0.22	0.28	0.40	-0.36
	C.I.	.004	.023	.020	.029	.037	
Drift correction using the exponential step distance D_{xs}							**ΔT.Cr**
*k*NN	Avg	0.71	0.74	0.53	0.53	0.51	+0.16
	C.I.	.013	.013	.012	.018	.021	
NNET	Avg	0.70	0.78	0.72	0.82	0.64	+0.19
	C.I.	.002	.006	.007	.022	.034	
PLS	**Avg**	**0.74**	**0.79**	**0.74**	**0.72**	**0.53**	**+0.31**
	C.I.	.011	.013	.014	.031	.029	
RF	Avg	0.88	0.77	0.81	0.81	0.90	+0.02
	C.I.	.004	.002	.003	.005	.019	

6.4 CMA-ES

The covariance matrix adaptation evolution strategy (CMA-ES) is an optimization method first proposed by Hansen, Ostermeier, and Gawelczyk [72] in mid 90s, and further developed in subsequent years [71], [70].

Similar to quasi-Newton methods, the CMA-ES is a second-order approach estimating a positive definite matrix within an iterative procedure. More precisely, it exploits a *covariance matrix*, closely related to the inverse Hessian on convex-quadratic functions. The approach is best suited for difficult non-linear, non-convex, and non-separable problems, of at least moderate dimensionality (i.e., $n \in [10, 100]$). In contrast to quasi-Newton methods, the CMA-ES does not use, nor approximate gradients, and does not even presume their existence. Thus, it can be used where derivative-based methods, e.g., *Broyden-Fletcher-Goldfarb-Shanno* or *conjugate gradient*, fail due to discontinuities, sharp bends, noise, local optima, etc.

In CMA-ES, iteration steps are called *generations* due to its biological foundations. The value of a generic algorithm parameter y during generation g is denoted with $y^{(g)}$. The mean vector $\mathbf{m}^{(g)} \in \mathbb{R}^n$ represents the favorite, most-promising solution so far. The *step size* $\sigma^{(g)} \in \mathbb{R}_+$ controls the step length, and the *covariance matrix* $\mathbf{C}^{(g)} \in \mathbb{R}^{n \times n}$ determines the shape of the distribution ellipsoid in the search space. Its goal is, loosely speaking, to fit the search distribution to the contour lines of the objective function f to be minimized. $\mathbf{C}^{(0)} = \mathbf{I}$

In each generation g, λ new solutions $\mathbf{x}_i^{(g+1)} \in \mathbb{R}^n$ are generated by sampling a multi-variate normal distribution $\mathcal{N}(\mathbf{0}, \mathbf{C})$ with mean $\mathbf{0}$ (see equation 6.14).

$$\mathbf{x}_k^{(g+1)} \sim \mathcal{N}\left(\mathbf{m}^{(g)}, \left(\sigma^{(g)}\right)^2 \mathbf{C}^{(g)}\right), k = 1, \ldots, \lambda \tag{6.14}$$

Where the symbol $\cdot \sim \cdot$ denotes the same distribution on the left and right side.

After the sampling phase, new solutions are evaluated and ranked. $\mathbf{x}_{i:\lambda}$ denotes the ith ranked solution point, such that $f(\mathbf{x}_{1:\lambda}) \leq \ldots \leq f(\mathbf{x}_{\lambda:\lambda})$. The μ best among the λ are selected and used for directing the next generation $g+1$. First, the distribution mean is updated (see equation 6.15).

$$\mathbf{m}^{(g+1)} = \sum_{i=1}^{\mu} w_i \mathbf{x}_i^{(g)}, w_1 \geq \ldots \geq w_\mu > 0, \sum_{i=1}^{\mu} w_i = 1 \tag{6.15}$$

In order to optimize its internal parameters, the CMA-ES tracks the so-called *evolution paths*, sequences of successive normalized steps over a number of generations. $\mathbf{p}_\sigma^{(g)} \in \mathbb{R}^n$ is the conjugate evolution path. $\mathbf{p}_\sigma^{(0)} = \mathbf{0}$. $\sqrt{2}\frac{\Gamma(\frac{n+1}{2})}{\Gamma(\frac{n}{2})} \approx \sqrt{n} + \mathcal{O}\left(\frac{1}{n}\right)$ is the expectation of the Euclidean norm of a $\mathcal{N}(\mathbf{0}, \mathbf{I})$ distributed random vector, used to normalize paths. $\mu_{\text{eff}} = \left(\sum_{1=1}^{\mu} w_i^2\right)^{-1}$ is usually denoted as *variance effective selection mass*. Let $c_\sigma < 1$ be the learning rate for cumulation for the rank-one update of the covariance matrix; $d_\sigma \approx 1$ be the damping parameter for step size update. Paths are updated according to equations 6.16 and 6.17.

$$\mathbf{p}_\sigma^{(g+1)} = (1-c_\sigma)\mathbf{p}_\sigma^{(g)} + \sqrt{c_\sigma(2-c_\sigma)\mu_{\text{eff}}}\mathbf{C}^{(g)-\frac{1}{2}}\frac{\mathbf{m}^{(g+1)}-\mathbf{m}^{(g)}}{\sigma^{(g)}} \tag{6.16}$$

$$\sigma^{(g+1)} = \sigma^{(g)}\exp\left(\frac{c_\sigma}{d_\sigma}\left(\frac{\left\|\mathbf{p}_\sigma^{(g+1)}\right\|}{\sqrt{2}\frac{\Gamma\left(\frac{n+1}{2}\right)}{\Gamma\left(\frac{n}{2}\right)}}-1\right)\right) \tag{6.17}$$

$\mathbf{p}_c^{(g)} \in \mathbb{R}^n$ is the evolution path, $\mathbf{p}_c^{(0)} = \mathbf{0}$. Let $c_c < 1$ be the learning rate for cumulation for the rank-one update of the covariance matrix. Let μ_{cov} be parameter for weighting between rank-one and rank-μ update, and $c_{\text{cov}} \leq 1$ be learning rate for the covariance matrix update. The covariance matrix $\mathbf{C}$ is updated (equations 6.18 and 6.19).

$$\mathbf{p}_c^{(g+1)} = (1-c_c)\mathbf{p}_c^{(g)} + \sqrt{c_c(2-c_c)\mu_{\text{eff}}}\frac{\mathbf{m}^{(g+1)}-\mathbf{m}^{(g)}}{\sigma^{(g)}} \tag{6.18}$$

$$\begin{aligned}\mathbf{C}^{(g+1)} &= (1-c_{\text{cov}})\mathbf{C}^{(g)} + \frac{c_{\text{cov}}}{\mu_{\text{cov}}}\\ &\times\left(\mathbf{p}_c^{(g+1)}\mathbf{p}_c^{(g+1)^T} + \delta\left(h_\sigma^{(g+1)}\right)\mathbf{C}^{(g)}\right)\\ &+c_{\text{cov}}\left(1-\frac{1}{\mu_{\text{cov}}}\right)\sum_{i=1}^{\mu} w_i \,\text{OP}\left(\frac{\mathbf{x}_{i:\lambda}^{(g+1)}-\mathbf{m}^{(g)}}{\sigma^{(g)}}\right)\end{aligned} \tag{6.19}$$

where $\text{OP}(\mathbf{X}) = \mathbf{X}\mathbf{X}^{\mathbf{T}} = \text{OP}(-\mathbf{X})$.

Most noticeably, the CMA-ES requires almost no parameter tuning for its application. The choice of strategy internal parameters is not left to the user, and even λ and μ defaults to acceptable values. Notably, the default population size λ is comparatively small to allow for fast convergence. Restarts with increasing population size has been demonstrated [8] useful for improving the global search performance, and it is nowadays included an an option in the standard algorithm.

6.5 Conclusions

In this chapter, we propose an evolutionary based approach able to deal with the drift problem affecting gas sensor arrays. The presented methodology is based on a 5-step flow that corrects and classifies the samples affected by sensor drift by applying a correction factor that mitigates the undesired effects on gas sensors. The correction factor is continuously adapted exploiting an evolutionary process, thus following the changes underwent by the sensor array due to the drift problem.

The proposed approach is flexible enough to work with different state-of-the-art classification algorithms, as experimentally demonstrated, and there is no need of relying upon complex drift models in order to exploit the proposed technique. Moreover, gathered results on artificial and real data sets experimentally corroborate that the proposed methodology performs better than state of the art methods, such as OSC.

Chapter 7
Development of On-Line Test Sets for Microprocessors

In software-based self-test (SBST) a microprocessor executes a set of test programs devised for detecting the highest possible percentage of faults. The main advantages of this approach are its high defect fault coverage (being performed at-speed) and the reduced cost (since it does not require any change in the processor hardware). SBST can also be used for on-line test of a microprocessor-based system. However, some additional constraints exist in this case (e.g. in terms of test length and duration, as well as intrusiveness). This paper faces the issue of automatically transforming a test set devised for manufacturing test in a test set suitable for on-line test. Experimental results are reported on an Intel 8051 microcontroller. Preliminary results have been published in [133].

7.1 Introduction

On-line testing has been defined as the process where faults are detected and/or corrected while the system is working in its natural environment. On-line test is required by most safety-critical applications, since a faulty behavior could lead to customers' inconveniences, economic loss and even casualties.

In *concurrent on-line testing* the detection of operational faults is performed at the same time the device is working. This means, more precisely, that the detection of operational faults must be performed keeping the system in normal or safety operational state. Concurrent online testing usually exploits specific hardware, such as data redundancy and voters, with a costly overhead for the final system.

Differently, in *non-concurrent on-line testing*, the detection of operational faults is performed while the normal operations are temporarily suspended [4]. The test procedure is either interruptible or composed of small subparts, and a test manager schedules it in order to minimize its intrusiveness. Non-concurrent testing usually needs less additional hardware. The development of an effective on-line test strategy is a major issue for designers of processor-based safety critical applications. Today, this task is becoming even more critical due to the growing number of

E. Sanchez et al.: Industrial Applications of Evolutionary Algorithms, ISRL 34, pp. 75–85.
springerlink.com

System-On-a-Chip's (SOCs) and other devices that embed microprocessor cores. Even though these systems are not yet used in mission-critical applications, their wide availability is often pushing producers to equip them with suitable solutions for on-line test.

Most non-critical applications are characterized by strict space/time constrains and high error latencies. The non-concurrent testing scheme may be preferred due to the limited hardware overhead required.

Several papers investigated different structures for on-line test. In [89], the authors propose the insertion of an *ad-hoc* test processor able to execute the on-line testing within the system. The processor has been developed and implemented looking for maximizing its performance regarding test; this means that internally the processor counts on special hardware structures such as Multiple Input Signature Registers (MISRs) and Linear Feedback Shift Registers (LFSRs) to improve test performance. The presented experience is mostly oriented to test the internal buses of the SOC. But the test processor could be empowered by additional subroutines able to test SOC peripherals and even functional processors. In [12] the main idea is to adopt an internal testing structure reusing the Infrastructure IPs (I-IPs), embedded into the SOC for manufacturing test to perform on-line testing. The strategy also includes a special I-IP named *test controller* to manage the test procedure. At the same time, several approaches have been developed for test program generation for manufacturing test of microprocessors (for instance [151], [32], [42] and [88]), exploiting the so-called *Software-Based Self-Test* (SBST) technique: the test of the processor is performed by letting it execute a suitable test program, whose results are analyzed in order to detect the existence of possible faults. This approach owns several advantages, including the fact that the test can be performed at-speed, and no special hardware is required within the processor. On the other side, the method effectiveness clearly depends on the quality of the adopted test program. In principle, the SBST approach could be extended to on-line test: a test set suitable for non-concurrent periodic on-line test could be composed of a certain number of small test programs. These test programs should cumulatively reach the target fault coverage, and they should be executed with a frequency suitable to guarantee the required fault latency and not to interfere with the normal functioning. The guidelines for writing an effective on-line test set may be found in [120]. Such test set may be composed of small programs that are activated during the idle periods of the system, or at specific time intervals (some programs may be even activated only at the system bootstrap and shutdown). If an operating system is available, the test can be handled by a single low-priority task that, each time is activated, launches a single test program and stores the result if accomplished without being interrupted, otherwise waits the next activation to relaunch the test. Such approach may only guarantee that on average the whole test is performed in a given amount of time, but performance degradation is negligible. On simpler systems, test programs can be handled by a scheduler. In any case, test programs should cumulatively guarantee the highest possible fault coverage, and should be characterized by

- a small code size, in order to reduce their impact in terms of memory occupation
- a short duration, in order to more easily fit into the available idle slots of the system (or to reduce their impact on the system performance)
- a minimal invasiveness, so that data variables used by the application are not affected by their execution.

Generally speaking, the number and size of test programs may be trade off for their intrusiveness.

This chapter presents a novel methodology that automatically generates a test set suitable for on-line test (according to the above requirements) starting from one devised for manufacturing test. The generated programs are suitable for non-concurrent periodic on-line test as well as for shutdown or startup testing, according to the constraints listed above. The generated test set could be applied to a microprocessor embedded into a SOC resorting to a test structure such as that described in [12], introducing low hardware overhead and avoiding performance degradation.

The chapter is organized as follows: section 2 describes the developed approach and section 3 presents a case study confirming the suitability of the approach. Finally, section 4 concludes the chapter and sketches future works.

7.2 Proposed Methodology

The proposed methodology is based on an automatic procedure which generates a test set suitable for a non-concurrent on-line testing starting from a conventional test set developed for post-production test. The approach is fully automated and guarantees to attain the same gate-level fault coverage (FC%) of the original test set.

The approach is based on an in-house developed instruction set simulator able to analyze existing test programs and split them up into a large number of fragments called *spores*. Each spore has some minimal test capability and owns the nice property of being small (in terms of code size and test duration). Then, using an evolutionary algorithm, the minimum set of such spores is chosen as the final test set. The proposed methodology is presented in the Figure 7.2. Remarkably, some older approaches like [45], exploited a structure called *macro*, very similar to spore, as a building block for constructing a test program. However, spores are the result of an analysis process, and not building blocks. While it may be maintained that every program can be broken into spores automatically, the opposite does not hold sensibly true. Later works showed that macros were inadequate to devise effective tests and proposed an enhanced structure based on graphs [88].

It must be noted that certain faults, like the ones affecting the address-related parts of the processor such as the logic that controls the program counter, may be detectable only by specific programs placed in specific memory locations. However, such tests are usually already small and fast, and therefore suitable for on-line test. On the contrary, the fault coverage is guaranteed on functional blocks, such as the multiplier unit, where the conversion is mostly needed.

In order to split the initial test set, each assembly program is executed by a special instruction set simulation (denoted as Spore Generator Program or SGP in Figure 7.2) able to infer each instruction data flow graph and to generate a small program able to thoroughly replicate the processor behavior while executing the referenced instruction. Each small program is named spore. Each spore represents a completely independent test program, able to excite some processor function, observe the results, and possibly signal fault occurrence. It is worth noting that the SGP module does not generate spores from scratch, but rather "learns" from the original test set how to test the processor modules, and extracts small groups of instructions, building a spore out of it.

Clearly, for each simulated program a huge set of spores is created. Then, using an evolutionary algorithm, the best set of programs able to guarantee at least the same FCattained by the original set is chosen as the final test set (On-line Test Set). Further information about the methodology is presented below.

The initial test set must guarantee a high FC%, but it does not need to be devised for online test. The method is independent on the origin of the initial test set, that could be generated by hand following some deterministic approach (as in [88]), or be the result of an automatic procedure (as in [42]), or even coming from a random generation. Additionally, functional programs used as specific tests to cover corner cases can also be used to increase the profits of the initial test set. Test programs for

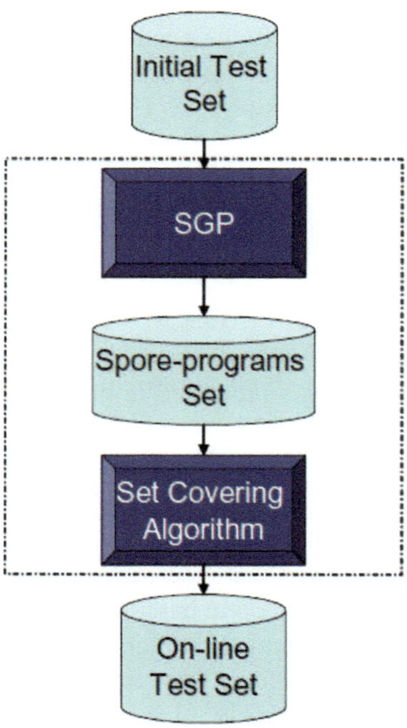

Fig. 7.1 General Methodology

post-production test are not devised to be executed sharing the processor resources; therefore, their effectiveness may be reduced if they are arbitrarily interrupted by task switching or other interruptions. Thus, they are incompatible with other applications working in background or, even worse, waiting for the processor to compute its tasks.

However, post-production test sets contain valuable information regarding FC%, and this information could be reused for the benefits of on-line testing. Finally, it must be noted that our method is independent on the way and frequency the online test is activated: our goal is only to transform the original test set into a new one which guarantees the same FC, but satisfies the requirements for on-line test, i.e., is composed of a minimal number of short test programs, which can be activated independently at different times.

The techniques used for activating the test and scheduling test programs depend specifically on the user requirements. All related issues, like the trade-off between test intrusiveness and error latency, are out of the scope of this work.

7.2.1 Spore Generator Description

The *Spore Generator Program* (SGP) is based on an instruction set simulator able to trace the execution data flow of each instruction of the test program. Its goal is the generation of independent and small programs able to exactly replicate the behavior of the processor while executing a target instruction.

To fit testing requirements, a spore program must be structured as shown in Figure 7.2.1. The program initialization regards controllability: the processor is set in a specific state waiting for the execution of the target instruction. At this point, the target instruction is executed and then appropriate instructions are added to make the results visible. These instructions depend on the chosen architecture and on the hardware exploited for on-line test, and are likely to be more restrictive than the observability points utilized in post-production tests.

To generate the set of spores, the Spore Generator uses the following components:

- The set of all elements required to emulate the current processor status, e.g., accumulators, flag registers and address registers
- The assembly syntax, addressing modes, and relationships between processor registers
- The mechanism used to observe results, such as moving the data to a specific processor output port or even just to a specific memory location.

The Spore Generator does not require to be time accurate.

An example of the approach operation is shown in Figure 7.2.1. A test program based on a loop is presented as initial test program. Then, the Spore Generator emulates its execution and for each simulated instruction (in this case the target instruction is `MUL AB`) a spore is generated. Resorting to the processor status and the data

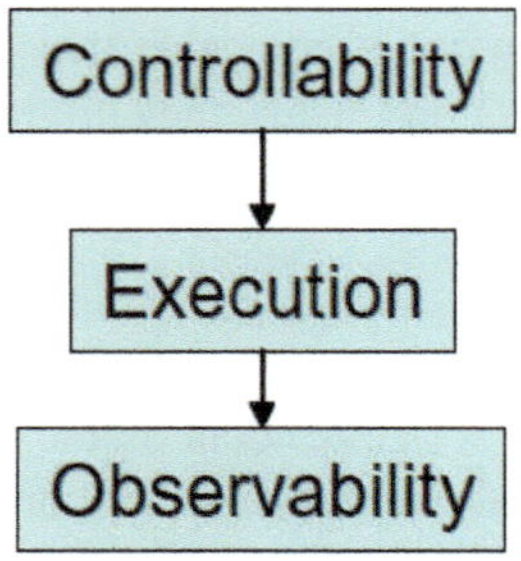

Fig. 7.2 Spore Structure

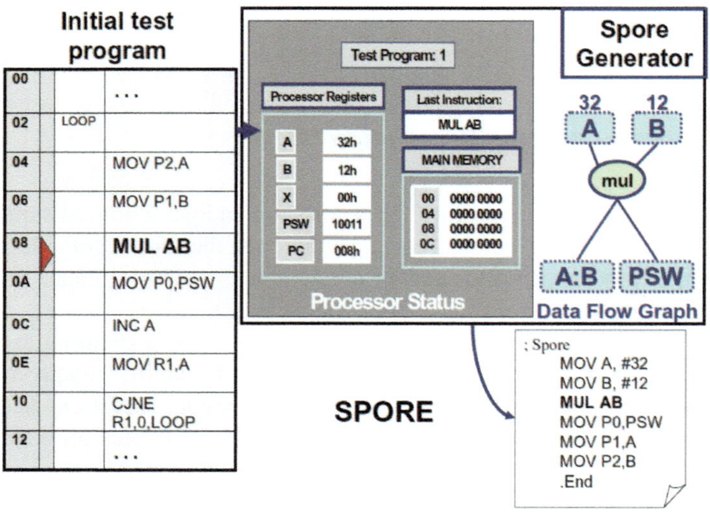

Fig. 7.3 Spore Generator, example.

flow graph the Spore Generator is able to create a spore that carefully emulates the processor behavior when the specific `MUL AB` instruction is executed.

To generate a spore set, it is important to properly configure the Spore Generator components expressing the particularities of the on-line testing scheme. For instance, the preceding example uses the processor ports P0, P1 and P2 to observe results. The spores set is able to achieve at least the same FC% the initial test program did, because the generated programs reflect step by step the behavior the processor underwent while it executed the initial test.

Despite the execution time required to compute the whole set of spores is larger than that required to execute the initial program, these programs are suitable for on-line testing because they comply with the characteristics mentioned before. However, the number makes their deployment prohibitive.

Therefore, it is necessary to rely on an automatic technique able to effectively choose a reduced set of test programs for on-line testing.

7.2.2 Set Covering

The problem of selecting a minimum sub set of the spores while still guaranteeing unmodified fault coverage is a mere set-covering, a well-known NP-hard problem for which several acceptable heuristics were developed in the past. The Selfish-Gene (SG) algorithm is an evolutionary heuristic algorithm developed in 1998 [46] and it was chosen mainly for its flexibility.

Differently from usual evolutionary algorithm, the SG uses the gene as unit of selection, and the selection mechanism guarantees gene correlations, i.e., the fact that a gene can be good or bad depending on the context of the other genes, exactly as the biological process does according to some researchers [48].

The Selfish Gene algorithm considers the whole population as a container of genetic material instead of storing a discrete set of individuals. At each step of the evolution process, two individuals are extracted from the population and their fitness is compared (fitness calculations are always performed at the level of individuals) simulating a conflict. Then, the population is updated to reflect the result of the battle.

From a practical point of view, the SG algorithm shares several advantages with common evolutionary algorithm, but since it does not store an explicit set of individuals, it is more easily scalable. Moreover, tuning the initial probabilities of the allele (a process called polarization), the behavior of the SG can be shaped from pure evolutionary to strongly hill climbing-like [44].

The SG was used to evolve a genome on n loci, where n is the number of spores. For each locus there are two possible alleles: 0 and 1, encoding that the test program is, respectively, included or not included in the on-line test set.

As final result, the set covering solver delivers a reduced set of programs able to achieve high FC% without large memory space requirements.

Finally, to manage this set of programs, a scheduler should be used. This scheduler must be additionally able to save and evaluate the testing signature properly to possibly stop the system operation in the case an error occurs. For instance, the on-line testing structure presented in [12] is appropriate for this task and does not introduce excessive hardware overheads.

7.3 Case Study

As a case study, we evaluated the presented methodology targeting an Intel i8051 processor. Only the processor primary output ports P0, P1 and P2 were chosen to observe results, making the on-line test strategy realistically applicable with limited hardware.

We exploited a synthetizable model of the processor described at RTL in about 14K VHDL code lines. The processor was synthesized using the *Synopsys Design Analyzer* and mapped on a standard library. Table 7.1 summarizes its main characteristics.

Table 7.1 Intel 8051 description

Primary Inputs	41
Primary Outputs	45
Gates	39,154
Flip-Flops	1,326
Stuck-at Faults	**72,672**

The processor core could be roughly divided in three units: Memory, Control and ALU. Table 7.2 shows the stuck-at faults (S@F) belonging to each part.

Table 7.2 Main processor parts

UNIT	S@F
Memory	43,960
Control	15,328
ALU	13,384

As it is well known, the test of the processor memory could be performed following a March test as described in [155]. Therefore, the experiments presented here target only the Control and ALU units of the processor, excluding the processor memory. The *Spore Generator* described previously was developed and implemented in about 3K lines of ANSI C code. The set covering algorithm has been implemented in about 1.5K lines of C code.

All the experiments have been performed on a Sun Enterprise 250 running at 400 MHz and equipped with 2 GBytes of RAM.

We considered an initial test set, developed in house and suitable for post-production testing, composed of 35 test programs. Five out of thirty five programs were devised following the methodology described in [88] and targeted the microprocessor Arithmetic and Logic Unit (ALU), while the remaining programs specifically targeted the Control Unit (CU).

Table 7.3 Initial test set figures

UNIT	#Prog	Size [bytes]	CC	FC%
ALU	5	220	136M	95.30
Control	30	4.3K	21K	90.1
TOTAL	**35**	**4.5K**	**136M**	**92.52**

Table 7.3 presents the figures obtained by the initial test set. The column "# Prog" shows the number of programs devised to tackle each unit; Column "Size" reports the size of the test programs in bytes; Column "CC" reports the number of clock cycles required to run the tests, and the final column labeled "FC%" reports the obtained fault coverage.

Stemming from the initial test set, two experiments were devised targeting the ALU and the CU separately. Experimental results are summarized in Table 7.4 for the ALU and in Table 7.5 for the CU.

On both of the tables, lines marked with I, S and F represent the Initial, Intermediate (Spores), and Final test sets, respectively. The column "#" shows the number of test programs, and as presented in the Table 7.3, figures regarding on program size and execution time are shown. In the block "Size [bytes]" the column "Max" shows the size of the longest test program, and the column "Tot" the cumulative size of the test set. Column block "CC" reports the number of clock cycles required to run the tests. Column "Max" shows the time to execute the longest test program and column "Tot" the time needed for running the whole test set. The column marked "FC%" details the fault coverage attained by each test set. The reader can see that the method was able to automatically generate a set of test programs attaining the same total fault coverage of the original one, with a significant reduction in the maximum size and duration of each test program, thus matching the requirements of on-line test.

Concerning the ALU experiments, the initial test set generates a huge number of spores cumulatively attaining the same fault coverage than the original test set on the target unit (95.3%). It must be observed that the final test set contains "only" 415 test programs (less than 0.1% of the full spore set) while still guaranteeing the same fault coverage. Compared with the initial test set, the on-line test set is larger in code size, but requires far less time to be completed. Moreover, it is composed of totally independent spores, none of them requiring more than 150 clock cycles to be run.

Table 7.4 ALU

	#	Size [bytes]		CC		FC%
		Max	Tot	Max	Tot	
I	5	53	220	40M	136M	95.3
S	6M	18	90M	150	178M	95.3
F	415	18	6,213	150	51,245	95.3

Table 7.5 Control Unit

	#	Size [bytes]		CC		FC%
		Max	Tot	Max	Tot	
I	30	160	4.3k	1k	21k	90.1
S	2.2k	47	66k	170	280k	83.2
F	250	47	7.5k	170	37k	83.2

In the case of the CU, the results obtained by the final test set report an increment in the size of the final test set as well as in the execution time and the data present a decrement in the fault coverage attained by the set. This reduction in the FC% can be explained by the fact that the spores are too small to fully stress the control logic regarding the address-related parts of the processor. However, it could be noted that, differently from the test programs targeting the ALU, the programs testing the CU do not need high quantities of CPU time to be executed. Thus, the final on-line test set for the whole microprocessor core could be composed of the test set generated for the ALU (415 spores), the test set for the CU (250 spores) and some of the initial programs targeting the CU, guaranteeing the same FC%. If time constrains are really strict, this additional content of the test set may be executed at startup or shutdown. As for the ALU, the obtained test programs show a significant reduction in their maximum size and duration.

The Spore Generator required few minutes to split the original test programs, and the selfish gene used few hours to optimize the test set. Interestingly, being based on an evolutionary approach, the user may reduce the computational effort lowering the quality of the result. The fault simulation of all the spores required about 10 days. Assessing the efficacy of all test programs is the most time consuming step of the approach, but it is an easily parallelizable task.

As a result, a test set devised for manufacturing test was automatically transformed in a new one suitable to be applied on-line. The initial test set is compact, requires a long time to be executed and is usually designed to be run without regarding sharing constraints. The final one is larger, but composed of small and extremely fast programs that can be freely scheduled. Both test sets guarantee the same fault coverage on the target units.

7.4 Conclusions

This chapter presented a fully automatic methodology able to transform a test set originally developed for manufacturing test in a test set suitable for on-line test. While the new test set is likely to contain a larger number of programs, these programs are shorter and completely independent (i.e., they can be executed at different times and do not rely each on the results of the previous ones), and thus perfectly fit a non-concurrent on-line test scheme.

The transformation of the test set is performed in two phases: first the original programs are simulated with a special instruction-set simulator that for each instruction generates a spore, i.e., a small program able to fully replicate the processor behavior. Second, an evolutionary algorithm is used to collapse the set of spores into a test set.

The proposed approach is able to guarantee the same fault coverage on all functional units. The experimental evaluation clearly shows the potentiality of the approach. Additionally, the time required to execute the final test set is lower than the one for the initial test set, leading to an improvement in the fault latency.

The approach may be extended to pipelined architectures: given a pipeline on *N* stages, each spore must load the required status and then execute the *N-1* instructions preceding the target one. As a result, the target instruction should be executed *exactly* in the same condition of the original one. Authors are currently extending the approach to deal with such microprocessors.

Part III
Test Generation Problems

Chapter 8
Uncovering Path Delay Faults with Multi-Objective EAs

This chapter presents an innovative approach for the generation of test programs detecting path-delay faults in microprocessors. The proposed method takes advantage of the multiobjective implementation of a previously devised evolutionary algorithm and exploits both gate- and RT-level descriptions of the processor: the former is used to build Binary Decision Diagrams (BDDs) for deriving fault excitation conditions; the latter is used for the automatic generation of test programs able to excite and propagate fault effects, based on a fast RTL simulation. Experiments on an 8-bit microcontroller show that the proposed method is able to generate suitable test programs more efficiently compared to existing approaches. Preliminary results have been published in [10].

8.1 Introduction

In order to guarantee product quality for today's microprocessor cores, traditional stuck-at tests are no longer sufficient and more complex fault models have to be considered when devising test strategies. At-speed delay fault testing, in particular, has been widely addressed by academia and is becoming common practice in industry [100][28][84]. Among all existing delay fault models, the path-delay fault model is considered the most accurate since it can detect both lumped and distributed delays [28][91], but also the most challenging, due to the enormous number of faults (paths). Delay test has been approached adopting different strategies, purely relying on an external tester or applying structural self-testing methodologies such as *Built-In Self-Test* (BIST), or exploiting the execution of suitable self-test programs. The latter strategy is usually referred to as *Software-Based Self-Test* (SBST) and is generally more affordable, as it exploits the processor instructions in the normal mode of operation; it can be used in stand-alone modules as well as when the processors are deeply embedded in a System on Chip (SoC) and their accessibility is reduced.

Regarding test generation addressing path-delay faults, several techniques exist for enhanced full-scan circuits, based on either structural ATPG tools [62][150] or

E. Sanchez et al.: Industrial Applications of Evolutionary Algorithms, ISRL 34, pp. 89–99.
springerlink.com

function-based tools using *Binary Decision Diagrams* (BDDs) [13][108][21] and Boolean-SAT [30][162] implementations. Some work on software-based test generation has been done exploiting deterministic techniques [141][94][67]. Evolutionary algorithms have been successfully exploited for the automatic generation of program sets for verification, test [43], and diagnosis [135] for processors described at different levels of abstraction. In most cases, the evolutionary algorithm faces the test set generation as a single-objective optimization problem, e.g., resorting to a multi-run strategy.

However, hardware optimization techniques belong to a real-world classification of problems that usually require the simultaneous optimization of many objectives. Therefore, hardware optimization problems could be addressed resorting to multiobjective optimizers. *Multiobjective Evolutionary Algorithms* (MOEAs) were initially introduced in 1985, by the implementation of the first evolutionary algorithm dealing with multiobjective optimization problems [138]. Roughly speaking, MOEAs produce a set of potentially optimal solutions, rather than an unique solution, that represents a subset of the Pareto optimal set.

This paper presents an innovative approach for the automatic generation of pathdelay functional test programs for microprocessors exploiting both gate- and RT-level descriptions. The former is used to select the set of critical paths to be considered and to obtain path excitation requirements based on BDD analysis; the latter is used for effectively identifying the test programs able to reproduce the conditions activating the targeted fault (*excitation*), and to make the fault effect(s) visible on the processor outputs (*propagation*). For automatically generating test programs, the new implementation of an evolutionary algorithm addressing multiobjective optimization is employed. The main advantage introduced is the improvement in the flow performances compared to other approaches based only on gate-level simulation [11].

The organization of this chapter is as follows: Section 2 provides the needed background; Section 3 details the proposed methodology; Section 4 presents the case of study. Finally, in Section 5 some conclusions are drawn.

8.2 Background

8.2.1 Software-Based Path Delay Testing

A path-delay fault occurs when a defect in a circuit causes the cumulative delay of a combinational path to exceed some specified duration [91][23]. The combinational path begins at a primary input or a clocked flip-flop (*startpoint*), includes a connected chain of gates, and ends at a primary output or a clocked flip-flop (*endpoint*) (Fig. 1). The specified time duration can be the duration of the clock period (or phase), or the vector period. The propagation delay is the time that an event (i.e., a transition) takes to traverse the path. For each combinational path in a circuit, there are two path-delay faults, corresponding to rising and falling transitions on the start-point. Signals that compose the path and feed the traversed gates are called *on-path signals*; signals that are not on the path but feed the gates on the path are called *off-path signals*. In order to examine the timing operation of a circuit we should examine signal transitions: delay tests consist of vector pairs ($V_1 \rightarrow V_2$) to be applied on the inputs feeding the path (*a*, *b*, *c*, *d* and *e* in Figure 8.2.1), so that an input transition on the startpoint propagates to the endpoint.

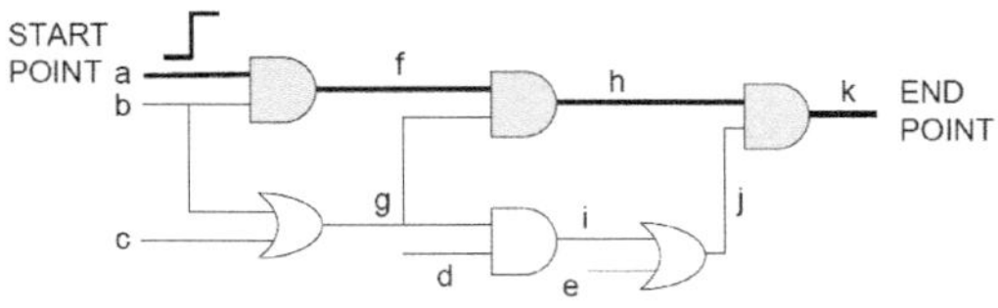

Fig. 8.1 Example of a path-delay fault: on-path signals indicated by thick lines (a, f, h, k); b, g and i are off-path signals.

Path-delay test application can be performed resorting to suitable scan-chains or by employing functional techniques. In scan-based test methodologies, the patterns are serially loaded into the scan chains (at reduced speed if necessary). Consequently, the two test vectors are applied in succession with a defined timing and the test results are shifted out through the scan chains, thus achieving full observability. In the case of Software-Based path-delay testing the test vectors $V_1 \rightarrow V_2$ reach the targeted path inputs during the normal at-speed circuit operations, hence depending on the sequence of data feed (instructions in case of processors) and allowing continuous application of test vectors. When targeting microprocessors, a test program must be made to ensure that the excitement conditions of the targeted path-delay fault are met in a consecutive pair of clock cycles, and that the fault effect(s) propagate to suitable observable points (e.g., output ports).

If a test can be applied in the normal operations of a circuit, we refer to it as a *functional test*. A path is *functionally testable* if there exists a functional test for that path. Otherwise, the path is *functionally untestable* [94]. Functionally untestable faults never determine the performance in normal operations of the circuit, and if detected during testing may lead to overkill (i.e., discarding functioning chips). On the other hand, defects on functional testable paths may degrade the circuit performance when path-delay faults occur. Software-based testing concentrates on the latter class, intrinsically avoiding over-testing redundant paths.

8.2.2 Exploiting Gate- and RT Level Descriptions for Path-Delay Testing

Commonly adopted solutions for path-delay test generation in sequential circuits are mostly based on the analysis of gate-level descriptions. Addressing a fault list provided by timing analysis tools, test patterns for path excitation are calculated. At this phase it is seldom possible to assess whether the faults are functionally testable. The test patterns correspond to two consecutive vectors to be applied at speed to the inputs of the combinational circuit partition including the selected path. From this point forward, they will be referred as V1 and V2.

When dealing with functional test (in the absence of scan structures) V1 and V2 are functionally justifiable iff they can be consecutively reproduced on the memory elements and primary inputs feeding the path by a sequence of instructions and data. In this case, the processor RT-level description may be employed to establish whether an instruction sequence is able to apply V1 and V2 to the selected combinational part. Since the observation of flip-flop values is required, only, it is possible to relate each considered flip-flop in the gate-level description to a signal in the RTL one.

8.2.3 BDDs for Structural Path Delay Fault Tests

Rather than devising a specific couple of vectors V1 and V2 that excite a specific fault, through BDD analysis of the gate-level netlist it is possible to derive a wider set of requirements for the combinational subcircuit inputs to excite the path it contains.

A reduced ordered Binary Decision Diagram (referred to as a BDD here) is a canonical graphical representation of a Boolean function [21]. BDDs have been widely used in test generation, for various fault models. For the case of path-delay faults in enhanced scan designs [13][108][116], given one (or more) fault(s) a Boolean function can be formulated whose solution space is all the possible pairs of test vectors that can detect the fault(s). This function is derived based on all the necessary values on on-path and off-path signals of the path-delay fault(s). The variables of the function correspond to the primary inputs of the circuit. When such a function is given by a BDD, we have a very compact (due to the suppression of variables with the x value) and implicit (non-enumerative) representation of the entire solution space. This is of high importance for several issues in test generation: untestable faults are very easily determined; hard-to-detect faults, that require a lot of time in structural-based ATPG tools, are also efficiently handled (BDD is very small since it contains a small number of cubes); fault simulation, for fault dropping, can be trivially performed on the BDD and not on the gate-level netlist. Moreover, if an input pattern is not a valid test, the BDD can be used to quickly determine how far the input pattern is from becoming a valid test (% of bits that must be changed in the input pattern). The latter is of particular importance in the proposed methodology, since it can quickly and accurately guide the evolutionary engine to generate the necessary path-delay fault tests.

8.2.4 Basic Concepts on MOEAs

Multiobjective evolutionary algorithms, as their single-objective counterpart, are population-based searching algorithms that mimic natural evolution. However, differently from single-objective algorithms, MOEAs exploit the population of individuals to simultaneously evolve solutions to multiple and usually conflicting goals [95][77]. The expected result from a MOEA is a set of trade-off individuals called nondominated solutions, Pareto-optimal solutions, or Pareto optimal set. For each individual into the population, a fitness vector $f_i = (x_1, x_2, ..., x_n)$ represents the figures of merit obtained by the individual regarding to the n pursued objectives.

Pareto optimality is defined using the concepts of domination: given two individuals A and B, A *dominates* B iff A is at least as good as B in all objectives, and better in at least one. A is *equivalent* to B iff results on A and B are identical in all objectives. A *covers* B if A either dominates or is equivalent to B. Similarly, given two sets of individuals Y and Z, Y dominates Z if every individual of Z is dominated by some individual of Y. Similar definitions relative to sets of individuals can be made for equivalence and coverage concepts. Thus, the Pareto optimal set is the set of all Pareto optimal individuals, and the corresponding set of fitness vectors is the *Pareto optimal front*. Individuals belonging to the Pareto optimal set are equally important. Indeed, for the individuals belonging to the Pareto optimal set, no improvement is possible in any objective without harming at least one of the other objectives.

Different strategies have been proposed in order to properly sort individuals belonging to the population; for example: aggregation-based approaches, lexicographical ordering, target-vector approaches, criterion-based approaches, and Pareto-based approaches. Some of them do not incorporate directly the concept of optimality outlined before, whereas others not only exploit it but include additional mechanisms to guarantee the diversity of the population. One of the most popular strategies used by MOEAs is based on a ranking scheme that divides the whole population on different sets, in such a way that each set contains only non-dominated individuals, and lower ranked sets are dominated by higher ones [95]. It is interesting to highlight that in a successful experiment the highest set contains the individuals belonging to the Pareto optimal set.

8.3 Proposed Approach

The proposed approach targets the automatic generation of test programs (i.e., instruction sequences) for processors addressing the path-delay fault model. This low-cost generation procedure exploits both gate- and RT-level descriptions.

Four main steps have been devised to approach the generation process:

Path list grouping A preliminary step, aimed at reducing the cost of the following generation step. The path list provided by timing analysis tools is analyzed and a set of shorter fault lists is produced, each one corresponding to a *coherent* set of critical paths in the processor netlist, i.e., a set of paths related to the same processor elements. As a matter of fact, excitation conditions for faults belonging to the same structurally coherent fault group are likely to be stressed by the same instructions. Details on this topic can be found in [11].

Circuit subdivision and BDD analysis Given the gate-level netlist and the addressed path list, for each path a combinational subcircuit (or *chunk*) is automatically extracted, which contains the path and, therefore, all the information needed for the analysis of its excitation conditions. A BDD is then derived that contains all the possible input vectors that bring necessary excitation values at the inputs of the path under consideration. Structurally untestable faults are removed in this phase. The BDD representation will be used in the *sequential fault excitation* step for evaluating the ability of each program to excite specific faults: the fitness function depends on the minimum hamming distance of the vectors applied from the set of vectors that can excite the path. It can be computed optimally and quickly when the set of vectors is represented by a BDD.

Sequential fault excitation This step aims at generating the test programs that effectively excite the considered path-delay faults. A MOEA is exploited to automatically generate instruction sequences, whose fitness is evaluated through RT-level simulation, avoiding highly expensive gate-level simulations, and relying on the already available BDDs. This step will be analyzed in detail.

Sequential Error propagation This step targets error propagation to the processor output ports and uses an evolutionary algorithm implementing a single-objective strategy. For this task, during the RTL simulation of the test program execution, the values of the flip-flops feeding the investigated path are analyzed at each clock cycle in order to check for the excitation conditions (both on on-path and off-path); whenever they are met, a faulty value is forced on the path endpoint for one clock cycle (*fault injection*, [11]). From that point in time, the state of all flip-flops is saved at each clock cycle and compared to the original (fault-free) simulation: if the simulation of the already generated program on the sabotaged RTL introduces a change on the processor output ports at any time following the fault injection, the test program achieves excitation and observation of the addressed fault and is complete. Otherwise, the number of flip-flops with different contents with respect to the fault-free simulation is used as a fitness function to be maximized, until the fault effects are propagated to the outputs.

The purpose of the sequential fault excitation phase (Figure 8.3) is the generation of suitable instruction sequences that excite the path-delay faults in coherent lists. This process is based on the usage of a new implementation of a well known *evolutionary algorithm* (EA), called μGP, able to automatically generate suitable test programs.

Roughly speaking, an EA is a population-based optimizer that imitates the natural process of biological evolution. Following this perspective, a test program is an *individual* and the tool handles a population of individuals (i.e., a collection of assembly programs). The initial population is generated randomly, then iteratively refined mimicking the Darwinian Theory: new individuals are generated either by *mutation* (an individual is slightly modified) or by *recombination* (two or more

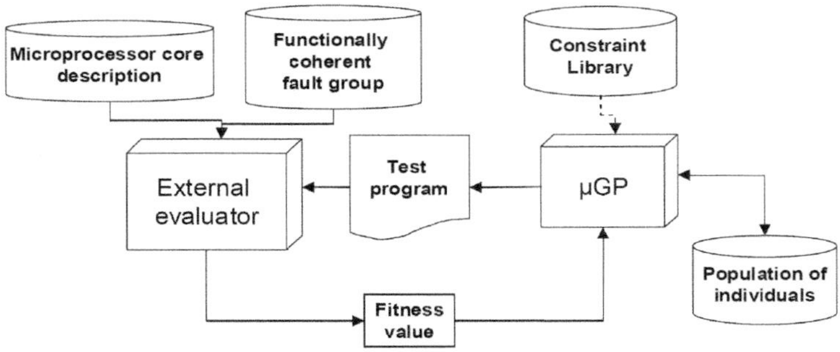

Fig. 8.2 Sequential fault excitation phase.

individuals are mixed in some way); the best performing individuals are selected for survival. The process is blocked after a certain number of steps, called *generations*, or when a steady state is reached. The best individual is eventually provided as output.

Differently from the standard approach described in [100], the evolutionary tool implements a MOEA [95] able to deal with several path-delay faults at a time. In this case the main goal of the evolutionary process is not to obtain a single best program but a set of best programs able to correctly excite the targeted faults. The main idea behind the MOEA implementation of μGP is to simultaneously optimize a complete functionally coherent group. As mentioned before, faults belonging to the same structurally coherent fault group are probably excited by similar test programs. Thus, the MOEA will evolve a population of individuals working on a specific portion of the processor core rather than a single program focusing on a unique fault.

μGP bases its evolutionary process on a constrained tagged graph, which is a directed graph whose elements may own one or more tags, and that in addition has to respect a set of constraints. The constraints may affect both the information contained in the graph elements and its structure. Graphs are initially generated in a random fashion; subsequently, they may be modified by genetic operators (e.g., the classical mutation and recombination, but also by different operators, as required; the tool architecture has been specially thought for easy addition of new genetic operators).

The purpose of the constraints is to limit the possible productions of the evolutionary tool, and also provide them with semantic value. The constraints are provided through a user-defined library that provides the genotype-phenotype mapping for the generated individuals, describes their possible structure and defines which values the existing parameters (if any) can take. Constraint definition is left to the user to increase the generality of the tool; it is flexible enough to allow the definition of complex entities to easily describe a wide range of processor *instruction sets architectures* (ISA).

The evolutionary core reads the *constraint library* in order to adequately generate assembly programs. For each generated program, a vector of fitness values are computed by the *external evaluator* considering the targeted faults provided by the *functionally coherent fault list*. Differently from the classical approach, the sequence of values in the fitness vector does not represent a priority list but each of them describes the figure of merit obtained by the individual regarding to a specific fault.

The task of the μGP core is to progressively improve the *population of individuals* or test programs. Thus, the population is ordered following a ranking strategy based on the Pareto-dominance principles described before. Choice of the individuals for reproduction is performed by means of a tournament selection based on the ranking position. However, since individuals belonging to the same group are by definition non-dominated ones, the selection is performed resorting to the *delta entropy value* of the individual [43]. The purpose of the entropy value is not to rank a population in absolute terms, but to detect whether the amount of genetic diversity in a set of individuals is increasing or decreasing. The tournament size £n is also endogenous. The population size μ is set at the beginning of a run, and the tool employs a variation on the plus $(\mu + \lambda)$ strategy: a configurable number λ of genetic operators are applied on the population. Since different operators may produce different number of offspring, the number of individuals added to the population is variable; the activation probability and strength for every operator is an endogenous parameter. All new unique individuals are then evaluated, and the population resulting from the union of old and new individuals is ordered resorting to the ranking approach described previously. Clearly, if a new individual dominates the complete population, a new individuals set is created and it is placed at the top of the rank list. Finally, only the first μ individuals are kept.

In order to customize this architecture to the specific goal we address here, we use the BDD-based *fitness function* described above, which is effective in guiding the algorithm towards the solution, and can be computed in reasonable times.

In this case, the evaluation of the generated test programs (or instruction sequences) is performed on the RT-level microprocessor core description by means of a logic simulation: during the simulation, at each clock cycle the vectors feeding the path are passed to the fitness function, and the maximum value obtained during the program run identifies the program's fitness.

8.4 Experimental Data

The proposed flow has been preliminary evaluated on a description of an 8051 microcontroller, addressing non-robust path-delay testing. The processor reads the test programs from an external memory and its output ports are directly accessible.

The critical timing analysis of the synthesized architecture has been performed utilizing the Synopsys PrimeTime suite ver. X-2005.12. The 92,430 worst paths were selected. This data is related to an in-house developed library. For each path, a combinational subcircuit is automatically extracted from the circuit and the BDD representation is generated and used to remove structurally untestable faults.

The set of structurally testable paths contains 10,394 faults. They have been automatically divided in classes depending on their structural coherence, using a simple tool based on set covering principles, and obtaining 96 coherent fault lists, each one including an average of about 108 faults.

The sequential fault excitation step has been performed resorting to the new MOEA implementation of μGP [100], which also includes a new operator called *local-scan mutation*, whose purpose is the generation of a reduced set of individuals in the neighborhood of the selected parent by performing slight mutations to only one determined parameter. In this case the fitness evaluator comprised a commercial logic simulator (Mentor Graphics ModelSim v.6.2h) and an ad-hoc C-language software monitor implemented in the simulator environment. The evolutionary experiment has been set up with the aim of performing a multi-objective optimization. The initial population is composed of 300 random individuals; the population size is 100 and at each generation 80 genetic operators are applied. For each of the coherent fault lists, the evolutionary experiment was set up in the following manner:

1. the first 20 faults in the list are initially considered (in order not to slow excessively the simulation, not all faults in the list are addressed together) and the EA is started, evaluating the excitation fitness (20 paths implies 20 fitness values);
2. whenever a test program fitness hits 100% for one of the inspected faults, that fault is removed from the experiment and replaced from a new one from the same list (fault dropping strategy). The obtained test program is saved;
3. if the algorithm does not improve the fitness for a set number of generations (10 in this case), the 20 paths are replaced with the following 20 in the list.

The process continues until all paths in the coherent fault list have been considered. This phase took about 110 hours for the whole fault list.

The error propagation step took about 35 hours. The fitness has been evaluated resorting to the ModelSim simulator running a script performing fault injection and to an ad-hoc tool elaborating the simulation dump. The majority of the test program set achieves test observability without modification; for the ones whose fault effects are still not propagated, the EA modifies the original test program maximizing the observability fitness, making sure that the excitation conditions are still met.

The obtained coverage values (Table 8.1) are comparable to the ones obtained using other approaches [94][11]. It must be noted that not-covered faults include functionally untestable ones, which do not determine the circuit performances and cannot be tested functionally. The required time computation compares favorably with the time required in [11]. The experiments run on an Intel E6400 @2.13 GHz.

In order to detail the behavior of the approach, the following pictures describe the evolution of an experiment targeting one coherent fault list that contains 84 faults. Figure 8.4 shows the first 300 steps of the evolutionary process: the continuous dark line represents the average of the 20 considered fitness values (mean value on the population), while vertical bars indicate the maximum fitness obtained at each step. For this coherent fault list, the final coverage is 50%. It is important to notice that whenever excitation is found for a fault (e.g., step 28), the average fitness falls down due to the fault dropping strategy. Similarly, this average value undergoes a big depression each time the steady state is reached and all targeted faults are replaced (steps 68, 118 and 189). Nevertheless, the average fitness tends to increase along the experiment. Figure 8.4 shows the first 50 steps of the same experiment; in this case, 5 out of the 20 evaluated fitness values are shown (average values on the population). Fitness 5 and 7 show that when a 100% is found the fitness value decreases, due to the substitution of the path-delay fault under inspection; however, the other fitness values seem not to be considerably affected by the replacement mechanism. It is also interesting to note that fitness 9 is continuously increased without finding a 100%. Finally, fitness 0 and 2 describe a very similar trajectory during the first 50

Table 8.1 Excitation and propagation figures on the case study.

	# of faults
Complete path set	92,430
Structurally Justified paths	10,394
Excited path-delay faults	2,731
Propagated faults (before error prop.)	1,536
Propagated faults (final)	2,489

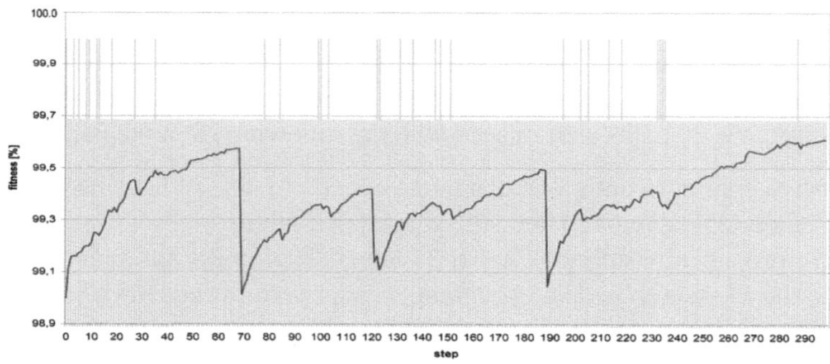

Fig. 8.3 Fitness behavior on a coherent path list, average and maximum values.

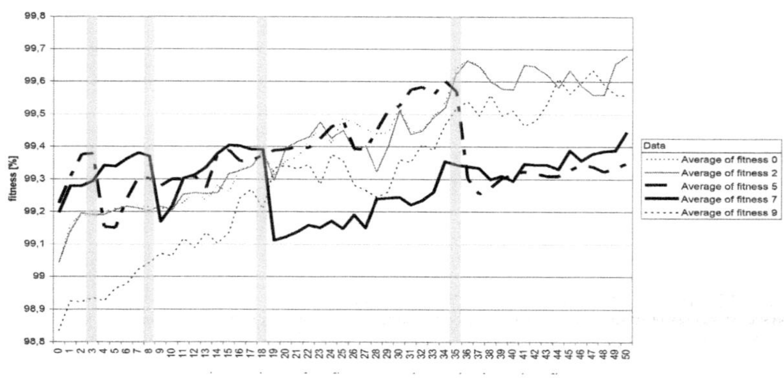

Fig. 8.4 Trajectories of 5 fitness values during the first 50 steps.

step, thus demonstrating the advantage of evolving coherent fault lists in the same experiment.

8.5 Conclusions

We presented an innovative approach to fully-automatic generation of path-delay test programs for microprocessors exploiting a MOEA.

Preliminary experimental results show that this methodology allows reducing the test generation time, by concentrating on suitably classified structurally coherent fault lists and avoiding computation-intensive gate-level simulations. The employed evolutionary algorithm takes advantage of the introduced BDD-based fitness evaluation functions for directing the test programs generation flow towards optimal solutions. The obtained coverage results are comparable to manual/deterministic approaches in literature.

Chapter 9
Software-Based Self Testing of System Peripherals

Traditional test generation methodologies for peripheral cores are performed by a skilled test engineer, leading to long generation times. In this paper a test generation methodology based on an evolutionary tool which exploits high level metrics is presented. To strengthen the correlation between high-level coverage and the gate-level fault coverage, in the case of peripheral cores, the FSMs embedded in the system are identified and then dynamically extracted via simulation, while transition coverage is used as a measure of how much the system is exercised. The results obtained by the evolutionary tool outperform those obtained by a skilled engineer on the same benchmark. Preliminary results have been published in [127].

9.1 Introduction

A system-on-chip (SoC) can integrate into a single device one or more processor cores with standard peripheral memory and application-oriented logic modules. This high integration of many components leads to an increased complexity of the test process since it decreases the accessibility of each functional module into the chip. Thus, the ever increasing usage of such devices demands for cheap testing methodologies.

The Software-based Self-test (SBST), whereby a program is executed on the processor core to extract information about the functioning of the processor or other SoC modules and provide it to the external test equipment [87] meets this demands since: it allows cheap at-speed testing of the SoC; it is relatively fast and flexible; it has very limited, if any, requirements in terms of additional hardware for the test; it is applicable even when the structure of a core is not known, or can not be modified. Even though SBST is currently being increasingly employed, the real challenge of software-based testing techniques is to generate effective test programs.

Many SBST techniques have been developed for the test of microprocessor cores; traditional methodologies resort to functional approaches based on exciting specific functions and resources of the processor [151]. New techniques, instead, differ on

E. Sanchez et al.: Industrial Applications of Evolutionary Algorithms, ISRL 34, pp. 101–110.
springerlink.com

the basis of the kind of description they start from: in some cases only the information coming from the processor functional descriptions are required [41]; other simulation-based approaches require a pre-synthesis RT-level description [33] or the gate-level description [39].

Simulation-based strategies are heavily time consuming, thus, the use of RT-level descriptions to drive the generation of test sets is preferable to allow much faster evaluation. Relying on high-level models not only helps the user of the SoC to perform more simulations increasing the confidence in the generated tests, but is also of value to the manufacturer allowing early generation of a significant part of the final test set. Whereas the correlation between RT-level code coverage metrics (CCM) and gate-level fault coverage is not guaranteed in the general case, several RT-level based methodologies maximize the CCMs to obtain a good degree of confidence on the quality of the generated test set.

This paper describes the application of an evolutionary algorithm in test set generation process for different types of peripheral cores embedded in a SoC. Furthermore the generation process is fully automated and requires a very low human effort. The generation process is driven by the transition coverage on the peripheral's finite state machine (FSM) and by the RT-level Code Coverage Metrics (CCMs).

Exploiting the correlation between high-level and low-level metrics, during the generation process only logic simulation is performed allowing the reduction of the generation time. The results are finally validated running a gate-level fault simulation.

Results show that the combination of the FSM transition coverage and CCMs can effectively guide the test block generation and a high fault coverage can be achieved. Moreover, we show that the new approach makes the test generation process more robust, improving the relationship between high- and low-level metrics.

The rest of the chapter is organized as follows: section 2 recalls some background concepts in peripheral testing; section 3 outlines the methodology adopted for the generation of test sets and details the evolutionary tool. Section 4 introduces the experimental setup, describing the case study and presents the experimental results. Finally, section 5 draws some conclusions.

9.2 Peripheral Testing

9.2.1 Basics

A typical SoC is composed of a microprocessor core, some peripheral components, memory modules, and possibly customized cores. An external ATE is supposed to be available for test application: its purpose is to load a test program in the memory, start execution, and interact with the peripherals applying data to the input ports and collecting values from the outputs while the program is running.

To make effective use of the test setup both the test programs and the peripheral input/output data have to be specified; therefore, a complete set for testing peripheral cores is composed of some test blocks [17], defined as basic test units composed of two parts: a configuration and a functional part. The configuration part includes a program fragment that defines the configuration modes used by the peripheral, and the functional part contains one or more program fragments that exercise the peripheral functionalities as well as the data set or stimuli set provided/read by the ATE.

Researchers have long sought high-level methodologies to generate high quality test sets; this is possible only if a correlation between high-level metrics and gatelevel fault coverage exists. Differently from the general case, where the correlation is vague, in the case of peripheral cores this correlation actually exists. It is not complete but, as experimentally shown in [15], suitable for test set generation.

Therefore, an automatic methodology for the generation of test sets for peripheral cores that uses a high-level model of the peripheral in the generation phase is an interesting solution to overcome new testing issues on SoCs.

As mentioned in [17], traditional code coverage metrics suitable for guiding the development of the test sets for peripheral cores are: Statement coverage (SC), Branch coverage (BC), Condition coverage (CC), Expression coverage (EC), Toggle coverage (TC). Maximizing all the coverage metrics allows to better exercise the peripheral core. It is not possible to accept a single coverage metric as the most reliable and complete one [98]; thus different metrics must be exploited in order to guarantee better performance of the test sets [144].

9.2.2 Previous Works

An attempt to provide effective solutions for peripheral test set generation is presented in [17]; the process is performed by hand and mainly relies on the experience of a test engineer, who maximizes sequentially the various coverage metrics, generating one or more test blocks for every metric. This process is repeated until sufficiently high coverage values are obtained for all the chosen metrics. In [80] a pseudo-exhaustive approach to generate functional programs for peripheral testing was presented. The proposed method generates a functional program for each possible operation mode of the peripheral core in order to generate control sequences which would place the peripheral in all possible functional modes. The pseudo-exhaustive approach produces a large number of functional programs, since one has to be written for every operation mode.

In [6] the authors describe a generic and systematic flow of SBST application on two communication peripheral cores. The methodology achieves high fault coverage but needs a deep knowledge of the peripheral core leading to long test development time with a high human effort. In [15] the peripheral test set generation has been automated using an evolutionary algorithm, called μGP.

The test block generation was supported by the construction of couples of templates: one for program and the other for data generation. The evolutionary algorithm is used to optimize parameter values, leaving the structure of the test block fixed. The obtained results compare favorably with respect to the manually generated [17].

In [16] an improved version of the evolutionary algorithm has been described, able to optimize both the structure and the parameters. The same results as [15] are obtained with no need of the rigid templates used previously, reducing significantly the required generation time.

9.3 Proposed Approach

As stated above, traditional CCMs extracted at the RT-level do not, in general, show a tight correlation with gate-level fault coverage. Furthermore, the RT-level descriptions use, especially in the case of complex cores, many modules that interact among each other in order to perform the core functionalities. The traditional CCMs do not consider these interactions and only aim at maximizing the coverage metrics in each module. After the synthesis process, at the gate level, the distinction between modules of a core is less clear and therefore it is important to consider the interactions to enforce a correlation between high-level metrics and low level ones.

One way to model a system is to represent it with a FSM. Coverage of all the possible transitions in the machine ensures thoroughly exercising the system functions. Additionally, the use of FSM transition coverage has the additional advantage that it makes the interactions between functional modules in the peripheral explicit. Figure 9.3 sketches the proposed methodology.

The evolutionary approach generates test blocks starting from information about the peripheral core and the processor assembly syntax only. Every new test block generated is evaluated using a high-level simulator. The evaluation stage assigns a fitness to every individual. The procedure ends when a time limit is elapsed or when a steady state is detected, that is, a predefined number of test blocks are generated without any improvement of the coverage metrics. At the end of the evolutionary run a single test block is provided as output.

The sketched procedure is iteratively repeated to generate a complete test set. In the steps following the first one, the evaluation phase is modified in order to only take into account the additional coverage provided by the new test blocks. The rationale for this methodology is that in general it is not possible to completely solve the problem with one single test block. The end result of the process is a set of test blocks that cumulatively maximize the targeted coverage metrics.

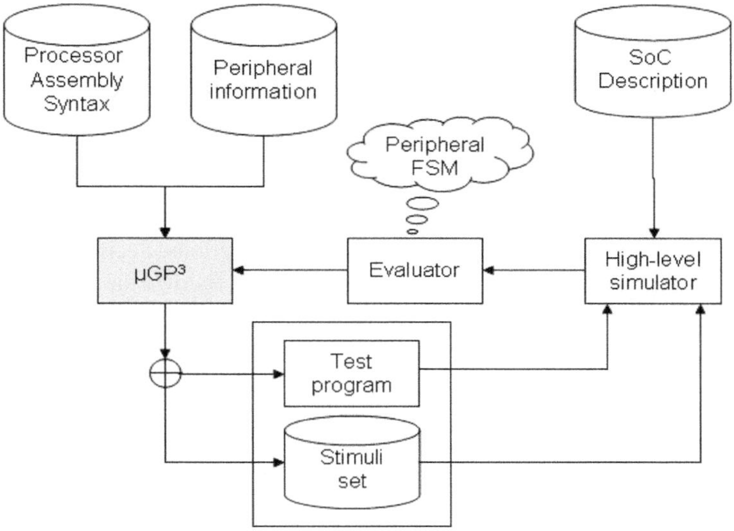

Fig. 9.1 Evolutionary generation loop.

9.3.1 Evolutionary Tool

For the automatic generation of the test blocks an evolutionary tool named ìGP3 [146] has been employed. μGP is a general-purpose approach to evolutionary computation, derived from a previous version specifically aimed at test program generation.

The tool is developed following the rules of software engineering and was implemented in C++. All input/output, except for the individuals to evaluate, is performed using XML with XSLT. The use of XML with XSLT for all input and output allows the use of standard tools, such as browsers, for inspection of the constraint library, the populations and the configuration options. The current version of the μGP comprises about 50,000 lines of C++ code, 113 classes, 149 header files and 170 C++ files.

Evolution Unit

μGP bases its evolutionary process on the concept of constrained tagged graph, that is a directed graph every element of which may own one or more tags, and that in addition has to respect a set of constraints. A tag is a name-value pair whose purpose is to convey additional information about the element to which it belongs, such as its name. Tags are used to add semantic information to graphs, augmenting the nodes with a number of parameters, and also to uniquely identify each element during the evolution. The constraints may affect both the information contained in the graph elements and its structure. Graphs are initially generated in a random fashion; subsequently, they may be modified by genetic operators, such as the classical mutation and recombination, but also by different operators, as required by the specific application. The tool architecture has been specially thought for easy addition of new genetic operators as needed by the application. The activation probability and strength for every operator is an endogenous parameter.

The genotype of every individual is described by one or more constrained tagged graphs, each of which is composed by one or more sections. Sections allow to define a global structure for the individuals that closely follows the structure of any candidate solution for the problem.

Constraints

The purpose of the constraints is to limit the possible productions of the evolutionary tool, and also provide them with semantic value. The constraints are provided through a user-defined library that provides the genotype-phenotype mapping for the generated individuals, describes their possible structure and to define which values the existing parameters (if any) can take.

Constraint definition is left to the user to increase the generality of the tool. The constraints are divided in sections, every section of the constraints matching a corresponding section in the individuals. Every section may also be composed of subsections and, finally, the subsections are composed of macros.

Constraint definition is flexible enough to allow the definition of complex entities, such as the test blocks described above, as individuals. Different sections in the constraints, and correspondingly in the individual, can map to different entities. In this specific case the constraints define three sections: a program configuration part, a program execution part and a data part or stimuli set. The first two are composed of assembly code, the third is written as part of a VHDL testbench. Though syntactically different, the three parts are interdependent in order to obtain good solutions. Fitness. Individual fitnesses are computed by means of an external evaluator: this may be any program able to provide the evolutionary core with proper feedback.

The fitness of an individual is represented by a sequence of floating point numbers optionally followed by a comment string. This is currently used in a prioritized fashion: one fitness A is considered greater than another fitness B if the n-th component of A is greater than the n-th component of B and all previous components (if any) are equal; if all components are equal then the two fitnesses are considered equal.

Evolutionary Scheme

The evolutionary tool is currently configured to cultivate all individuals in a single panmictic population, although it can be configured to use an island model. The population is ordered by fitness. Choice of the individuals for reproduction is performed by means of a tournament selection; the tournament size τ is also endogenous. The population size ì is set at the beginning of a run, and the tool employs a variation on the plus $(\mu + \lambda)$ strategy: a configurable number λ of genetic operators are applied on the population. Since different operators may produce different number of offspring the number of individuals added to the population is variable. All new unique individuals are then evaluated, and the population resulting from the union of old and new individuals is sorted by decreasing fitness. Finally, only the first μ individuals are kept.

To promote diversity, the individuals genetically equal to already existing ones, called clones, may have their fitness scaled by a fixed value in the range [0.0,1.0]. The possible termination conditions for the evolutionary run are: a target fitness value is achieved by the best individual; no fitness increase is registered for a predefined number of generations; a maximum number of generations is reached.

At the end of every generation the internal state of the algorithm is saved in a XML file for subsequent analysis and for providing a minimal tolerance to system crashes.

9.3.2 Evaluator

The proposed approach is based on modeling the entire system as a FSM which is dynamically constructed during the test generation process. Thus, differently from other approaches, the FSM extraction is fully automated, and requires minimum human effort: the approach only requires the designer to identify the state registers in the RT-level code; every global state in the peripheral represents a possible configuration of values of all the state registers. Thus, whenever a state register in any module changes its value, also the global state of the peripheral is affected. Given the dynamic nature of the FSM construction, it is not possible to assume known the maximum number of reachable states, not to mention the possible transitions. For this reason it is impossible to determine the transition coverage with respect to the entire FSM.

As experimentally demonstrated [98], maximizing more than one metric usually leads to better quality tests. Thereby, the simulation-based method proposed here exploits the FSM transition coverage, that enforce a maximum interaction between peripheral modules, and all the available CCMs to thoroughly exercise the peripheral functionalities.

The implemented evaluator collects the output of the simulation and dynamically explores the FSM; it assesses the quality of the test block considering the transition coverage on the FSM and the CCMs.

The fitness fed back to the evolutionary tool is composed of many parts: the FSM transition coverage followed by all the others CCMs (SC, BC, CC, EC, TC). As we mentioned before the metrics are considered in order of importance. In this way it is possible, during the generation process, to select more thoroughly those test blocks that are able to better excite the peripheral.

9.4 Experimental Analysis

9.4.1 Test Case

The benchmark is a purposely designed SoC which includes a Motorola 6809 microprocessor, a Universal Asynchronous Receive and Transmit (UART), a Peripheral Interface Adapter (PIA), a Video display unit (VDU) and a RAM memory core. The system derives from one available on an open source site [112]. The methodology is used to test the UART, the PIA and the VDU in the targeted SoC.

The peripherals are described at RT-level in VHDL code and are composed of different modules. The SoC was synthesized using a generic home-developed library.

Table 9.1 Implementation characteristics

Description	Measure	PIA	VDU	UART
RT-level	statements	149	153	383
	branches	134	66	182
	condition	75	24	73
	expression	0	9	54
	toggle	77	199	203
Gate level	Gates	1,016	1,321	2,247
	Faults	1,938	2334	4,054

Table 9.1 shows details of the targeted peripherals, including information at high and low level. Rows labeled with RT-level present CCM information while the remaining rows illustrate the number of gates counted on the synthesized devices and the number of collapsed faults for the stuck-at model, respectively.

At the end of the generation process, some gate-level fault simulation were performed only to validate the proposed methodology; the gate-level fault coverage figures reported in the following sections target the single stuck-at fault model.

9.4.2 Experimental Results

All the reported experiments have been performed on a PC with an Athlon XP3000 processor, 1GB of RAM, running Linux.

The algorithm parameters for the evolutionary experiments are the same both when targeting only the CCMs, and when the number of transitions in the FSM is also taken into account: for the PIA and the VDU experiments, $\mu = 50$ and $\lambda = 70$; and as the UART is more complex than the PIA the evolutionary parameters were set to perform a lower number of simulations: μ was set to 30 and λ to 40.

In order to provide the reader with a reference value, we recall that the fault coverage obtained by the manual approach presented in [17] is 80.96% for the UART and 89.78% for the PIA.

Table 9.2 summarizes the results obtained for the targeted peripherals, reporting the number of FSM transitions covered, the high-level CCMs and the stuck-at fault coverage (FC) in percentage. The reader should note that the value of traditional CCMs are expressed as absolute values (instead of percentages).

Table 9.2 Results for considered peripherals

	PIA	VDU	UART
FSM Transition	115	191,022	142
Statement	149	153	383
Branch	129	66	180
Condition	68	23	72
Expression	0	9	51
Toggle	77	191	203
FC(%)	91.4	90.8	91.28

For every peripheral considered the methodology is able to reach a good value of gate-level fault coverage. In the case of the VDU the number of transition is very high; this is due to the state registers that hold the current position on the screen.

To experimentally demonstrate that the use of the FSM transition coverage is essential to strengthen the correlation between high an low level metrics 100 experiments on the UART are performed, using both the evolutionary approach presented in [16] and the generation process detailed above.

Table 9.3 Comparison between the two methodologies

		FSM	SC	BC	CC	EC	TC	FC
[16]	Average	NA	381.8	178.7	70.7	50.7	201.3	84.8
	std.dev.	NA	0.36	0.39	0.30	0.32	0.40	6.37
New methodology	Average	141.0	382.2	179.3	71.8	50.8	202.2	90.9
	std.dev.	1.49	0.28	0.33	0.22	0.24	0.36	1.10

Table 9.3 reports a comparison between the results of the experiments performed following the methodology presented in [16] and the current one; the table illustrates

the average and standard deviation of the different CCMs and of the stuck-at fault coverage (FC). In all cases the CCMs are very near to the absolute maximum, and both methodologies lead to small standard deviations on the considered metrics. In the first case, however, the standard deviation in the fault coverage of each test set is relatively high. Although the methodology obtains good results, it is not as robust as desirable, and the obtained solution may not exhibit the expected quality.

Using the new methodology the average fault coverage is increased by more than 6% and, more importantly, the standard deviation of the fault coverage is dramatically reduced. This clearly shows that the robustness of the methodology is increased, and solutions of consistent quality can be obtained.

Table 9.4 Comparison between the two methodologies

	FC	**TGEN**	**TAPP**	**Size**
[16]	90.7	5.1	28,842	1,953/72
New Methodology	91.3	2.2	32,762	2,345/87

Table 9.4 synthetically reports a comparison between the two methodologies in the case of the UART, highlighting the obtained fault coverage (FC) in percentage, the average generation time (TGEN) expressed in hours, the average application time (TAPP) in clock cycles, and the average size of the test sets, reported as program bytes and data bytes. The results clearly show that the new methodology outperforms the previous one in terms of fault coverage and generation time. The latter, in particular, is less than a half with respect to the previous methodology, highlighting the efficiency of the new approach.

Other approaches [80][6] to peripheral test are not directly comparable with our methodology since they are referred to different devices, although their complexity and the results are similar to the devices analyzed here. Furthermore, our methodology only needs RT-level simulation and does not need the time-expensive fault-simulations.

9.5 Conclusions

In this chapter, a successful application of the evolutionary tool for the generation of sets of test blocks for different types of peripheral modules in SoCs driven by the FSM transition coverage and the high-level CCM has been described.

The evolutionary tool is able to generate test blocks where the relation between high-level coverage metrics and low level one is much stronger; this better relation has been experimentally demonstrated with a experimental analysis where many test blocks are generated and evaluated.

The experimental results on different type of peripheral cores, communication peripherals and VDU controller, show the effectiveness of the proposed methodology.

Chapter 10
Software-Based Self-Testing on Microprocessors

Microprocessor testing is becoming a challenging task, due to the increasing complexity of modern architectures. Nowadays, most architectures are tackled with a combination of scan chains and Software-Based Self-Test (SBST) methodologies. Among SBST techniques, evolutionary feedback-based ones prove effective in microprocessor testing: their main disadvantage, however, is the considerable time required to generate suitable test programs. A novel evolutionary-based approach, able to appreciably reduce the generation time, is presented. The proposed method exploits a high-level representation of the architecture under test and a dynamically built Finite State Machine (FSM) model to assess fault coverage without resorting to time-expensive simulations on low-level models. Experimental results, performed on an OpenRISC processor, show that the resulting test obtains a nearly complete fault coverage against the targeted fault model.

Results of this work have been accepted for publication in [136].

10.1 Introduction

In the last years, the market demand for a higher computational performance in embedded devices has been continuously increasing for a wide range of application areas, from entertainment (smart phones, portable game consoles), to professional equipment (palmtops, digital cameras), to control systems in various fields (automotive, industry, telecommunications). The largest part of today's Systems-on- Chip (SoCs) includes at least one processor core. Companies have been pushing design houses and semiconductor producers to increase microprocessor speed and computational power while reducing costs and power consumption. The performance of processor and microprocessor cores has impressively increased due to technological and architectural aspects. Microprocessor cores are following the same trend of high-end microprocessors and quite complex units may be easily found in modern SoCs.

E. Sanchez et al.: Industrial Applications of Evolutionary Algorithms, ISRL 34, pp. 111–120.
springerlink.com

Technology advancements impose new challenges to microprocessor testing: as device geometries shrink, deep-submicron delay defects are becoming more prominent [100], thereby increasing the need for at-speed tests; as core operating frequency and speed of I/O interfaces rise, more expensive external test equipment is required.

The increasing size and complexity of microprocessor architectures directly reflects in more demanding test generation and application strategies. Modern designs contain intricate architectures that increase test complexity. Indeed, pipelined and superscalar designs demonstrated to be random pattern resistant [42]. The use of hardware-based approaches, such as scan chains and BIST, even though consolidated in industry for integrated digital circuits, has proven to be often inadequate, since these techniques introduce excessive area overhead [22], require extreme power dissipation during the test application [158], and are often ineffective when testing delay-related faults [147].

As a consequence, academy is looking for novel paradigms to respond to the new testing issues: one promising alternative to hardware-based approaches is to exploit the processor to execute carefully crafted test programs. The goal of these test programs is to uncover possible design or production flaws in the processor. This technique, called Software-Based Self-Test (SBST), has been already used in different problems with positive results.

In this paper, we propose a SBST simulation-based framework for the generation of post-production test programs for pipelined processors. The main novelty of the proposed approach is its ability to efficiently generate test programs, exploiting a high level description of the processor under test, while the evolution of the generation is driven by the transition coverage of a FSM created during the evolution process itself.

10.2 Background

10.2.1 Software-Based Self Testing

SBST techniques have many advantages over other testing methodologies, thanks to their features: the testing procedure can be conducted with very limited area overhead, if any; the average power dissipation is comparable with the one observable in mission mode; the possibility of damages due to excessive switching activity, non-negligible in other methods, is virtually eliminated; test programs can be run at the maximum system speed, thus allowing testing of a larger set of defects, including delay-related ones; the approach is applicable even when the structure of a module is not known or cannot be modified.

SBST approaches proposed in literature do not necessarily aim to substitute other established testing approaches (e.g., scan chains or BIST) but rather to supplement them by adding more test quality at a low cost. The objective is to create a test program able to run on the target microprocessor and test its modules, satisfying the target fault coverage requirements. Achieving this test quality requires a proper test program generation phase, which is the main focus of most SBST approaches in recent literature. The quality of a test program is measured by its coverage of the design errors or production defects, its code size, and time required for its execution. The available approaches for test program generation can be classified according to the processor representation that is employed in the flow. High-level representations of the processor Instruction Set Architecture (ISA) or state transition graphs are convenient for limiting the complexity of the architecture analysis, and provide direct correlation with specific instruction sequences, but cannot guarantee the detection of structural faults. Lower-level representations, such as RT and gate-level netlists, describe in greater detail the target device and allow to focus on structural fault models, but involve additional computational effort.

A survey on some of the most important techniques developed for test program generation is presented in [125]. Due to modern microprocessors' complex architectures, automatic test program generation is a challenging task: considering different architectural paradigms, from pipelined to multithreaded processors, the search space to be explored is even larger than that of classic processors. Thus, it becomes crucial to devise methods able to automate as much as possible the generation process, reducing the need for skilled (and expensive) human intervention, and guaranteeing an unbiased coverage of corner cases.

Generation techniques can be classified in two main groups: *formal* and *simulation-based*. Formal methodologies exploit mathematical techniques to prove specific properties, such as the absence of deadlocks or the equivalence between two descriptions. Such a proof implicitly considers all possible inputs and all accessible states of the circuit. Differently, simulation-based techniques rely on the simulation of a set of stimuli to unearth misbehaviors in the device under test. A simulation-based approach may therefore be able to demonstrate the presence of a bug, but will never be able to prove its absence: however, the user may assume that the bug does not exist with level of confidence related to the quality of the simulated test set. The generation of a qualifying test set is the key problem with simulation-based techniques. Different methodologies may be used to add contents to such test sets, ranging from deterministic to pseudo-random.

Theoretically, formal techniques are able to guarantee their results, while simulation-based approaches can never reach complete confidence. However, the former require considerable computational power, and therefore may not be able to provide results for a complex design. Moreover, formal methodologies are routinely applied to simplified models, or used with simplified boundaries conditions. Thus, the model used could contain some differences with respect to the original design, introducing a certain amount of uncertainty in the process [124].

Nowadays, simulation-based techniques dominate the test generation arena for microprocessors, with formal methods bounded to very specific components in the earliest stages of the design. In most of the cases, simulation-based techniques exploit *feedback* to iteratively improve a test set in order to maximize a given target measure. Nevertheless, simulation of low-level descriptions could require enormous efforts in terms of computational time, memory occupation and hardware.

The main drawback of feedback-based simulation methods, is the long elaboration time required during the evolution. When dealing with a complete processor core, for example in [144], the generation time increases when low abstraction descriptions are used as part of the evolution: the growth of computation times is mainly due to the inclusion of *fault simulation* in the process. For every possible fault of the design, the original circuit is modified including the considered fault; then, a complete simulation is performed in order to understand whether the fault changes the circuit outputs. Even though lots of efforts are spent on improving this process [104], several minutes are still required to perform a fault simulation on a processor core with about 20k faults.

10.2.2 Evolutionary Algorithms on Software-Based Self Testing

Several approaches that face test program generation by exploiting an automated methodology have been presented in recent years: in [81] a tool named VERTIS, able to generate both test and verification programs based on the processor's instruction set architecture only, is proposed. VERTIS generates many different instruction sequences for every possible instruction being tested, thus leading to very large test programs. The test program generated for the GL85 processor following this approach is compared with the patterns generated by two Automatic Test Pattern Generator (ATPG) tools: the test program achieves a 90.20% stuck-at fault coverage, much higher than the fault coverage of the ATPG tools, proving the efficacy of SBST for the first time. The VERTIS tool works with either pseudo-random instruction sequences and random data, or with test instruction sequences and heuristics to assign values to instruction operands specified by the user in order to achieve good results. In more complex processors, devising such heuristics is obviously a non-trivial task.

In [119], an automated functional self-test method, called Functional Random Instruction Testing at Speed (FRITS), is presented. FRITS is based on the generation of random instruction sequences with pseudorandom data. The authors determine the basic requirements for the application of a cache-resident approach: the processor must incorporate a cache load mechanism for the test program downloading and the loaded test program must not produce either cache misses or bus cycles. The authors report some results on an Intel Pentium©4 processor: test programs automatically generated by the FRITS tool achieve 70% stuck-at fault cover-age for the entire chip, and when these programs are enhanced with manually generated tests, the fault coverage increases by 5%.

Differently from the previously described functional methods, in [45] the authors propose a two-steps methodology based on evolutionary algorithms: firstly a set of macros encrypting processor instructions is created, and in a second step an evolutionary optimizer is exploited to select macros and data values to conform the test program. The proposed approach is evaluated on a synthesized version of an 8051 microprocessor, achieving about 86% fault coverage. Later, in [42], a new version of the proposed approach is presented. The authors exploit a simulation-based method that makes use of a feedback evaluation to improve the quality of test programs: the approach is based on an evolutionary algorithm and it is capable of evolving small test programs that capture target corner cases for design validation purposes. The effectiveness of the approach is demonstrated by comparing it with a pure instruction randomizer, on a RTL description of the LEON2 processor. With respect to the purely random method, the proposed approach is able to seize three additional intricate corner cases while saturating the available addressed code coverage metrics. The developed validation programs are more effective and smaller in code size.

10.3 Proposed Approach

The previously described test generation cases show that evolutionary algorithms can effectively face real-world problems. However, when exploiting a low-level description of the processor under evaluation, simulation-based approaches require huge elaboration times.

We propose a methodology able to exploit a high-level description of a pipelined processor core in the generation process: the required generation time is thus reduced with respect to techniques that use a low-level description during the generation phase, such as the gate-level netlist, as reported in [144]. In the proposed approach, it must be noticed that the processor netlist is only used at the end of the generation process to assess the methodology results, performing a complete fault simulation. The generation process is supported by the on-time automated generation of a FSM that models the excited parts of the processor core and drives the evolution process by indicating the unreached components on the processor core. In addition, we consider also high-level coverage metrics to improve the evolution.

It is possible to simply define a pipelined microprocessor as the interleaving of sequential elements (data, state and control registers), and combinational logic blocks. The inputs of the internal combinatory logic blocks are dependent on the instruction sequence that is executed by the processor and on the data that are processed. One way to model a microprocessor is to represent it with a FSM. Coverage of all the possible transitions in the machine ensures thoroughly exercising the system functions. Additionally, the use of the FSM transition coverage has the additional advantage that it explicitly shows the interactions between different pipeline stages. Thus, we define the state word of a pipelined processor FSM model as the union of all logic values present in the sequential elements of the pipeline, excluding only the values strictly related to the data path. Consequently, the FSM transits to a new state at every clock cycle, because at least one bit in the state word is changed due to the whole interaction of the processor pipeline.

Figure refframework shows the proposed framework. The evolutionary core, called ìGP3 [146], is able to generate syntactically correct assembly programs by acquiring information about the processor under evaluation from an user-defined file called Constraint Library. When the process starts, the evolutionary core generates an initial set of random programs, or individuals, exploiting the information provided by the library of constraint. Then, these individuals are cultivated following the Darwinian concepts of natural evolution. Every test program is evaluated resorting to external tools that simulate the high level description of the processor core, resorting to a logic simulator at RTL, and generate a set of high-level measures. Contemporary, during the logic simulation, the FSM status is captured at every clock cycle, and for every evaluated test program the visited states and the traveled transitions are reported back to the evolutionary core as part of the evaluation of the goodness of an individual, called fitness value. The interaction between the different elements composing the fitness value guarantees good quality regarding the fault coverage against a specific fault model at gate level. Fitness values gathered during the logic simulation, for example code coverage metrics such as Statement coverage (SC), Branch coverage (BC), Condition coverage (CC), Expression coverage (EC), Toggle coverage (TC), are suitable for guiding the evolution of test programs. Simultaneously, maximizing the number of traversed transitions of the FSM model, assures a better result at gate level.

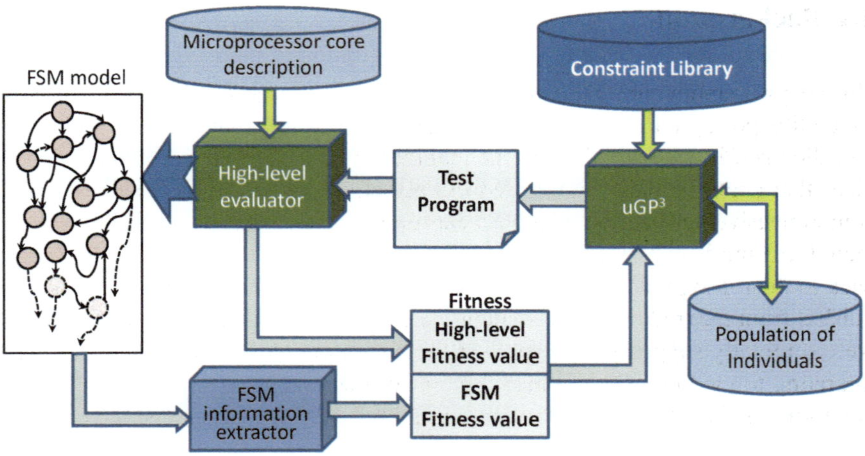

Fig. 10.1 Test generation framework.

Only the best individual is fault simulated in order to assess its fault coverage properties, reducing generation times. In the following paragraphs, we briefly describe in more detail the most significant elements present in the described framework.

10.3.1 μGP

μGP represent individuals, in this case candidate test programs, as constrained tagged graphs; a tagged graph is a directed graph every element of which may own one or more tags, and that in addition has to respect a set of constraints. A tag is a namevalue pair used to add semantic information to graphs, augmenting the nodes with a number of parameters, and also to uniquely identify each element during the evolution. Graphs are initially generated in a random fashion; subsequently, they may be modified by genetic operators, such as the classical mutation and recombination. The genotype of an individual is described by one or more constrained tagged graphs.

The purpose of the constraints is to limit the possible productions of the evolutionary tool, also providing them with semantic value. The constraints are provided through a user-defined library that supplies the genotype-phenotype mapping for the generated individuals, describes their possible structure and defines which values the existing parameters (if any) can assume. To increase the generality of the tool, constraint definition is left to the user.

In this specific case the constraints define three distinct sections in an individual: a program configuration part, a program execution part and a data part or stimuli set. The first two are composed of assembly code, the third is written as part of a VHDL testbench. Though syntactically different, the three parts are interdependent in order to obtain good solutions.

Individual fitness values are computed by means of one or more external evaluator tools. The fitness of an individual is represented by a sequence of floating point numbers optionally followed by a comment string. This is currently used in a prioritized fashion: one fitness A is considered greater than another fitness B if the nth component of A is greater than the n-th component of B and all previous components (if any) are equal; if all components are equal then the two fitnesses are considered equal.

The evolutionary tool is currently configured to cultivate all individuals in a single panmictic population. The population is ordered by fitness. Choice of the individuals for reproduction is performed by means of a tournament selection; the tournament size τ is also endogenous. The population size μ is set at the beginning of a run, and the tool employs a variation on the plus $(\mu + \lambda)$ strategy: a configurable number λ of genetic operators is applied on the population. All new unique individuals are then evaluated, and the population resulting from the union of old and new individuals is sorted by decreasing fitness. Finally, only the first μ individuals are kept.

The possible termination conditions for the evolutionary run are: a target fitness value is achieved by the best individual; no fitness increase is registered for a predefined number of generations; a maximum number of generations is reached.

10.3.2 FSM Extractor

The proposed methodology is based on modeling the entire processor core as a FSM which is dynamically constructed during the test generation process. Thus, differently from other approaches, the FSM extraction is fully automated, and demands minimum human effort: the approach only requires the designer to identify the memory elements of the pipeline registers in the RTL processor description that will determine state characteristics of the FSM. The key point behind the FSM extractor is to guide the evolution trough a high-level model of the processor core that summarizes the capacity of excitation of the considered test program. The FSM information extractor receives from the external evaluator (e.g., a logic simulator) the activity of the pipeline registers of the processor core at every clock cycle, then, it computes for every clock cycle the processor state word and extracts the visited states and the traversed transitions.

Given the dynamic nature of the FSM construction, it is not possible to assume as known the maximum number of reachable states, not to mention the possible transitions. For this reason, it is impossible to determine the transition coverage with respect to the entire FSM.

The implemented evaluator, that includes the logic simulator and the FSM information extractor, collects the output of the simulation and dynamically explores the FSM; it assesses the quality of test program considering the transition coverage on the FSM and the code coverage metrics. The fitness fed back to the evolutionary tool

is composed of many parts: the FSM transition coverage followed by all other highlevel metrics (SC, BC, CC, EC, TC).

Let us consider the mechanisms related to hazard detection and forwarding activation in a pipelined processor: in order to thoroughly test them, it requires to stimulate the processor core with special sequences of strongly dependent instructions able to activate and propagate possible faults on these pipelined mechanisms. Facing this problem by hand requires a very good knowledge about the processor core to carefully craft a sequence of instructions able to actually excite the mentioned pipelined elements. Additionally, this process may involve a huge quantity of time. On the other hand, state-of-the-art test programs usually do not target such pipeline mechanisms, since their main concern is exciting a targeted functional unit through carefully selected values, and not to activate the different forwarding paths and other mechanisms devoted to handle data dependency between instructions [126].

As a matter of fact, it is possible to state that a feedback based approach able to collect information about the interaction of the different instructions in a pipelined processor as the one described before, allows the evolution of sequences of dependent instructions that excite the mentioned pipeline mechanisms.

10.4 Case Study and Experimental Results

The effectiveness of the EA-based proposed methodology has been experimentally evaluated on a benchmark SoC that contains the OpenRISC processor core and some peripheral cores, such as the VGA interface, PS/2 interface, Audio interface, UART, Ethernet and JTAG Debug interface. The SoC uses a 32 bit WISHBONE bus rev. B for the communication between the cores. The operating frequency of the SoC is 150 MHz. The implemented SoC is based on a version publicly available at [112].

The OpenRISC processor is a 32 bit scalar RISC architecture with Harvard microarchitecture, 5 stages integer pipeline and virtual memory support. It includes supplementary functionalities, such as programmable interrupt controller, power management unit and high-resolution tick timer facility. The processor implements a 8Kbyte data cache and a 8Kbyte instruction cache 1-way direct mapped; the instruction cache is separated from the data cache because of the specifics of the Harvard microarchitecture.

In our experiments we decide to tackle specifically the processor integer unit (IU) that includes the whole processor pipeline. This unit is particularly complex and important in pipelined processors, since it is in charge of handling the flow of instructions elaborated in the processor core.

The pipelined processor is described by eight verilog files, counting about 4,500 lines of code. Table 1 describes some figures that are used to compute RTL code coverage and toggle metrics. Additionally, the final line shows the number of stuck-at faults (*S@ faults*) present in the synthesized version of the targeted module. The state word is defined as the union of all memory elements composing the processor pipeline, excluding only the registers that contain data elements. Data registers are excluded because we are mainly interested in the control part of the pipeline, and not in the data path.

Table 10.1 Details of the Integer Unit.

OR1200 IU	
Lines	4,546
Statements	466
Branches	443
Condition	53
Expression	123
Toggle	3,184
S@ faults	13,248

Thus, considering the registers available in every stage of the processor pipeline, a state word contained 237 bits is saved at every clock cycle during the logic simulation of a test program allowing us to dynamically extract the processor FSM. In order to monitor the elements contained in the state word of the pipeline at every clock cycle, we implemented a Programming Language Interface module, called

PLI, that captures the information required during the logic simulation of the RTL processor. The PLI module is implemented in C language, counting about 200 lines of code. The module is compiled together with the RTL description of the processor core, exploiting a verilog wrapper.

Once a test program is simulated, a PERL script extracts the information regarding the number of visited states as well as the number of traversed transitions obtained by the considered program. This information is collected together to the high-level coverage metrics provided by the logic simulator and the complete set of values is fed back to the evolutionary engine in the form of fitness value of the test program. The configuration files for the evolutionary optimizer are prepared in XML and count about 1,000 lines of code. Finally, additional perl scripts are devised to close the generation loop.

A complete experiment targeting the OR1200 pipeline requires about 5 days. At the end of the experiment, an individual counting 3,994 assembly lines that almost saturate the high level metrics is created; the same individual obtains about 92% fault coverage against the targeted fault model.

Compared to manual approaches reported in [126], that achieve about 90% fault coverage in the considered module, the results obtained in this paper improve the fault coverage by about 2%, and can be thus considered promising.

With the specifications given above, a complete FSM could theoretically comprehend a maximum of 2^{237} states. It is interesting to note that the final FSM actually counts only about 10^4, and despite this, the fault coverage is extremely high. This interesting result will be explored to gain further information on the layout of the complete FSM.

The final FSM could also be exploited to analyze the behavior of the device from a functional point of view, and it could be used to integrate the developed test program: deterministic programs could be created to reach the states untouched by the execution of the final best individual in the population.

Appendix A
References

1. Great internet mersenne prime search, http://www.mersenne.org/
2. Intel math kernel library, http://software.intel.com/en-us/intel-mkl/
3. Richards Adrion, W., Branstad, M.A., Cherniavsky, J.C.: Validation, verification, and testing of computer software. ACM Comput. Surv. 14, 159–192 (1982)
4. Al-Asaad, H., Murray, B.T., Hayes, J.P.: Online bist for embedded systems. IEEE Design Test of Computers 15(4), 17–24 (1998)
5. Aliwell, S.R., Halsall, J.F., Pratt, K.F.E., O'Sullivan, J., Jones, R.L., Cox, R.A., Utembe, S.R., Hansford, G.M., Williams, D.E.: Ozone sensors based on wo3: a model for sensor drift and a measurement correction method. Measurement Science & Technology 12(6), 684–690 (2001)
6. Apostolakis, A., Psarakis, M., Gizopoulos, D., Paschalis, A.: A functional self-test approach for peripheral cores in processor-based socs. In: 13th IEEE International On-Line Testing Symposium, IOLTS 2007, pp. 271–276 (2007)
7. Artursson, T., Eklov, T., Lundström, I., Mårtensson, P., Sjöström, M., Holmberg, M.: Drift correction for gas sensors using multivariate methods. Journal of Chemometrics, Special Issue: Proceedings of the SSC6 14(5-6), 711–723 (1999)
8. Auger, A., Hansen, N.: A restart cma evolution strategy with increasing population size. In: Proc. IEEE Congress Evolutionary Computation, vol. 2, pp. 1769–1776 (2005)
9. Baeck, T., Fogel, D.B., Michalewicz, Z. (eds.): Handbook of Evolutionary Computation. IOP Press (1997)
10. Bernardi, P., Christou, K., Grosso, M., Michael, M.K., Sánchez, E., Reorda, M.S.: Exploiting MOEA to Automatically Geneate Test Programs for Path-Delay Faults in Microprocessors. In: Giacobini, M., Brabazon, A., Cagnoni, S., Di Caro, G.A., Drechsler, R., Ekárt, A., Esparcia-Alcázar, A.I., Farooq, M., Fink, A., McCormack, J., O'Neill, M., Romero, J., Rothlauf, F., Squillero, G., Uyar, A.Ş., Yang, S. (eds.) EvoWorkshops 2008. LNCS, vol. 4974, pp. 224–234. Springer, Heidelberg (2008); 10.1007/978-3-540-78761-7_23
11. Bernardi, P., Grosso, M., Sanchez, E., Reorda, M.S.: On the automatic generation of test programs for path-delay faults in microprocessor cores. In: 12th IEEE European Test Symposium, ETS 2007, pp. 179–184 (May 2007)
12. Bernardi, P., Rebaudengo, M., Reorda, M.S.: Exploiting an i-ip for in-field soc test. In: 19th IEEE International Symposium on Defect and Fault Tolerance in VLSI Systems, DFT 2004, pp. 404–412 (2004)

13. Bhattacharya, D., Agrawal, P., Agrawal, V.D.: Test generation for path delay faults using binary decision diagrams. IEEE Transactions on Computers 44(3), 434–447 (1995)
14. Bhattacharyya, A.B.: Compact MOSEFT Models for VLSI Design. Wiley Publishing (2008)
15. Bolzani, L., Sanchez, E., Schillaci, M., Reorda, M.S., Squillero, G.: An automated methodology for cogeneration of test blocks for peripheral cores. In: 13th IEEE International On-Line Testing Symposium, IOLTS 2007, pp. 265–270 (2007)
16. Bolzani, L., Sanchez, E., Schillaci, M., Squillero, G.: Co-evolution of test programs and stimuli vectors for testing of embedded peripheral cores. In: IEEE Congress on Evolutionary Computation, CEC 2007, pp. 3474–3481 (2007)
17. Bolzani, L.M.V., Sanchez, E.E., Reorda, M.S.: A software-based methodology for the generation of peripheral test sets based on high-level descriptions. In: Proceedings of the 20th Annual Conference on Integrated Circuits and Systems Design, SBCCI 2007, pp. 348–353. ACM, New York (2007)
18. Box, G.E.P.: Evolutionary operation: A method for increasing industrial prouctivity. Applied Statistics VI(2), 81–101 (1957)
19. Brause, R., Langsdorf, T., Hepp, M.: Neural data mining for credit card fraud detection. In: 11th IEEE International Conference on Tools with Artificial Intelligence (1999)
20. Bremermann, H.J.: Optimization through Evolution and Recombination. Spartan Books (1962)
21. Bryant, R.E.: Graph-based algorithms for boolean function manipulation. IEEE Transactions on Computers C-35(8), 677–691 (1986)
22. Bushard, L., Chelstrom, N., Ferguson, S., Keller, B.: Dft of the cell processor and its impact on eda test software. In: Asian Test Symposium, pp. 369–374 (2006)
23. Bushnell, M.L., Agrawal, V.D.: Essentials of Electronic Testing for Digital, Memory and Mixed-Signal VLSI Circuits. Kluwer, Boston (2000)
24. Cagnoni, S., Dobrzeniecki, A.B., Poli, R., Yanch, J.C.: Genetic algorithm-based interactive segmentation of 3d medical images. Image and Vision Computing 17(12), 881–895 (1999)
25. Callegari, N., Wang, L.-C., Bastani, P.: Speedpath analysis based on hypothesis pruning and ranking. In: Proceedings of the 46th Annual Design Automation Conference, DAC 2009, pp. 346–351. ACM, New York (2009)
26. Cannon, W.D.: The Wisdom of the body. W.W.Norton (1932)
27. Chakraborty, A., Duraisami, K., Sathanur, A., Sithambaram, P., Benini, L., Macii, A., Macii, E., Poncino, M.: Dynamic thermal clock skew compensation using tunable delay buffers. IEEE Transactions on Very Large Scale Integration (VLSI) Systems 16(6), 639–649 (2008)
28. Chakraborty, T.J., Agrawal, V.D., Bushnell, M.L.: Delay fault models and test generation for random logic sequential circuits. In: Proceedings of 29th ACM/IEEE Design Automation Conference 1992, pp. 165–172 (June 1992)
29. Chellapilla, K., Hoorfar, A.: Evolutionary programming: an efficient alternative to genetic algorithms for electromagnetic optimization problems. In: IEEE Antennas and Propagation Society International Symposium 1998, vol. 1, pp. 42–45 (June 1998)
30. Chen, C.-A., Gupta, S.K.: A satisfiability-based test generator for path delay faults in combinational circuits. In: Proceedings of 33rd Design Automation Conference 1996, pp. 209–214 (June 1996)
31. Chen, D.Y., Chan, P.K.: An intelligent isfet sensory system with temperature and drift compensation for long-term monitoring. IEEE Sensors Journal 8(12), 1948–1959 (2008)

32. Chen, L., Dey, S.: Defuse: a deterministic functional self-test methodology for processors. In: Proceedings of 18th IEEE VLSI Test Symposium 2000, pp. 255–262 (2000)
33. Cheng, C., Lim, C.-C., Parashkevov, A.: A software test program generator for verifying system-on-chips. In: Tenth IEEE International on High-Level Design Validation and Test Workshop 2005, pp. 79–86 (2005)
34. Choi, S., Sarkar, T.K., Choi, J.: Adaptive antenna array for direction-of-arrival estimation utilizing the conjugate gradient method. Signal Processing 45(3), 313–327 (1995)
35. Christou, K., Michael, M.K., Bernardi, P., Grosso, M., Sanchez, E., Reorda, M.S.: A novel sbst generation technique for path-delay faults in microprocessors exploiting gate- and rt-level descriptions. In: 26th IEEE VLSI Test Symposium, VTS 2008, April 27-May 1, pp. 389–394 (2008)
36. Cinque, M., Cotroneo, D., Kalbarczyk, Z., Iyer, R.K.: How do mobile phones fail? a failure data analysis of symbian os smart phones. In: 37th Annual IEEE/IFIP International Conference on Dependable Systems and Networks, DSN 2007, pp. 585–594 (2007)
37. Collet, P., Lutton, E., Raynal, F., Schoenauer, M.: Polar ifs+parisian genetic programming=efficient ifs inverse problem solving. Genetic Programming and Evolvable Machines 1, 339–361 (2000), doi:10.1023/A:1010065123132
38. Colwell, B.: The zen of overclocking. Computer 37(3), 9–12 (2004)
39. Corno, F., Cumani, G., Sonza Reorda, M., Squillero, G.: An rt-level fault model with high gate level correlation. In: Proceedings of IEEE International on High-Level Design Validation and Test Workshop, pp. 3–8 (2000)
40. Corno, F., Cumani, G., Sonza Reorda, M., Squillero, G.: Efficient machine-code test-program induction. In: Proceedings of the 2002 Congress on Evolutionary Computation, CEC 2002, vol. 2, pp. 1486–1491 (2002)
41. Corno, F., Cumani, G., Sonza Reorda, M., Squillero, G.: Fully automatic test program generation for microprocessor cores. In: Design, Automation and Test in Europe Conference and Exhibition 2003, pp. 1006–1011 (2003)
42. Corno, F., Sanchez, E., Reorda, M.S., Squillero, G.: Automatic test program generation: a case study. IEEE Design Test of Computers 21(2), 102–109 (2004)
43. Corno, F., Sanchez, E., Squillero, G.: Evolving assembly programs: How games help microprocessor validation. IEEE Transactions on Evolutionary Computation, Special Issue on Evolutionary Computation and Games 9, 695–706 (2005)
44. Corno, F., Sonza Reorda, M., Squillero, G.: A New Evolutionary Paradigm for Cultivating Cellular Automata for Built-In Self Test of Sequential Circuits. In: Evolutionary Algorithms for Embedded System Design, pp. 143–173. Kluwer Academic Publishers (2002)
45. Corno, F., Sonza Reorda, M., Squillero, G., Violante, M.: On the test of microprocessor ip cores. In: Proceedings of Design, Automation and Test in Europe, Conference and Exhibition 2001, pp. 209–213 (2001)
46. Corno, F., Sonza Reorda, M., Squillero, G.: The selfish gene algorithm: a new evolutionary optimization strategy. In: Proceedings of the ACM Symposium on Applied Computing, SAC 1998, pp. 349–355. ACM, New York (1999)
47. Darwin, C.: On the origin of species, ch. IV, p. 502. John Murray (1859)
48. Dawkins, R.: The Selfish Gene. Oxford University Press (1976)
49. Di Carlo, S., Falasconi, M., Sánchez, E., Scionti, A., Squillero, G., Tonda, A.: Exploiting Evolution for an Adaptive Drift-Robust Classifier in Chemical Sensing. In: Di Chio, C., Cagnoni, S., Cotta, C., Ebner, M., Ekárt, A., Esparcia-Alcazar, A.I., Goh, C.-K., Merelo, J.J., Neri, F., Preuß, M., Togelius, J., Yannakakis, G.N. (eds.) EvoApplicatons 2010. LNCS, vol. 6024, pp. 412–421. Springer, Heidelberg (2010)

50. D'Silva, V., Kroening, D., Weissenbacher, G.: A survey of automated techniques for formal software verification. IEEE Transactions on Computer-Aided Design of Integrated Circuits and Systems 27(7), 1165–1178 (2008)
51. Duda, R.O., Hart, P.E., Stork, D.G.: Pattern Classification, 2nd edn. Wiley Interscience (2000)
52. El-Far, K.I., Whittaker, J.A.: Model-Based Software Testing. In: Encyclopedia of Software Engineering. Wiley-Interscience, New York (1994)
53. Elliot, R.S. (ed.): Antenna theory and design. Prentice-Hall, Inc. (1981)
54. Falasconi, M., Gutierrez, A., Pardo, M., Sberveglieri, G., Marco, S.: A stability based validity method for fuzzy clustering. Pattern Recogn. 43(4), 1292–1305 (2010)
55. Flautner, K., Patel, D.I.: Intelligent energy management^{TM} for portable embedded systems. In: Proceedings of IEEE International Conference on SOC, p. 415 (2003)
56. Fogel, D.B.: Evolutionary computation: toward a new philosophy of machine intelligence. IEEE Press, Piscataway (1995)
57. Fogel, L.J.: Autonomous automata. Industrial Research 4, 14–19 (1962)
58. Fogel, L.J.: Toward inductive inference automata. In: Proceeding of the International Federation for Information Processing Congress, pp. 395–400 (1962)
59. Frazer, A.S.: Simulation of genetic systems by automatic digital computers (part 1). Australian Journal of Biological Science 10, 484–491 (1957)
60. Frazer, A.S.: Simulation of genetic systems by automatic digital computers (part 1). Australian Journal of Biological Science 10, 492–499 (1957)
61. Friedberg, R.M.: A learning machine: Part i. IBM Journal 2(1), 2–13 (1958)
62. Fuchs, K., Pabst, M., Rossel, T.: Resist: a recursive test pattern generation algorithm for path delay faults considering various test classes. IEEE Transactions on Computer-Aided Design of Integrated Circuits and Systems 13(12), 1550–1562 (1994)
63. Gandini, S., Ravotto, D., Ruzzarin, W., Sanchez, E., Squillero, G., Tonda, A.: Automatic detection of software defects: an industrial experience. In: Proceedings of the 11th Annual Conference on Genetic and Evolutionary Computation, GECCO 2009, pp. 1921–1922. ACM, New York (2009)
64. Gandini, S., Ruzzarin, W., Sanchez, E., Squillero, G., Tonda, A.: A framework for automated detection of power-related software errors in industrial verification processes. Journal of Electronic Testing, 1–9 (2010); doi:10.1007/s10836-010-5184-5
65. Gobbi, E., Falasconi, M., Concina, I., Mantero, G., Bianchi, F., Mattarozzi, M., Musci, M., Sberveglieri, G.: Electronic nose and alicyclobacillus spp. spoilage of fruit juices: An emerging diagnostic tool. Food Control 21(10), 1374–1382 (2010)
66. Gould, S.J.: The Dinosaur in the Haystack. Harmony Books (1995)
67. Gurumurthy, S., Vemu, R., Abraham, J.A., Saab, D.G.: Automatic generation of instructions to robustly test delay defects in processors. In: 12th IEEE European Test Symposium, ETS 2007, pp. 173–178 (May 2007)
68. Gutierrez-Osuna, R.: Drift reduction for metal-oxide sensor arrays using canonical correlation regression and partial least squares. In: Proceedings of the 7th International Symp. on Olfaction and Electronic Nose, July 20-24, p. 147. Institute of Physics Publishing (2000)
69. Hamlet, D.: Random Testing. In: Encyclopedia of Software Engineering. Wiley-Interscience, New York (1994)
70. Hansen, N., Müller, S.D., Petrosnf, P.K.: Reducing the time complexity of the derandomized evolution strategy with covariance matrix adaptation (CMA-ES). Evolutionary Computation 11, 1–18 (2003)
71. Hansen, N., Ostermeier, A.: Completely derandomized self-adaptation in evolution strategies. Evolutionary Computation 9, 159–195 (2001)

72. Hansen, N., Ostermeier, A., Gawelczyk, A.: On the adaptation of arbitrary normal mutation distributions in evolution strategies: The generating set adaptation. In: Proceedings 6th International Conference on Genetic Algorithms, pp. 312–317. Morgan Kaufmann (1995)
73. Haugen, J.-E., Tomic, O., Kvaal, K.: A calibration method for handling the temporal drift of solid state gas-sensors. Analytica Chimica Acta 407(1-2), 23–39 (2000)
74. Hines, E.L., Llobet, E., Gardner, J.W.: Electronic noses: a review of signal processing techniques. IEEE Proceedings Circuits, Devices and Systems 146(6), 297–310 (1999)
75. Holland, J.H.: Adaptation in natural and artificial systems. MIT Press, Cambridge (1992)
76. Hoorfar, A., Zhu, J.: A novel hybrid ep-ga method for efficient electromagnetics optimization. In: IEEE Antennas and Propagation Society International Symposium, vol. 1, pp. 310–313 (2002)
77. Huband, S., Hingston, P., Barone, L., While, L.: A review of multiobjective test problems and a scalable test problem toolkit. IEEE Transactions on Evolutionary Computation 10(5), 477–506 (2006)
78. Ding, H., Liu, J.-H., Shen, Z.-R.: Drift reduction of gas sensor by wavelet and principal component analysis. Sensors and Actuators B: Chemical 96(1-2), 354–363 (2003)
79. Ionescu, R., Vancu, A., Tomescu, A.: Time-dependent humidity calibration for drift corrections in electronic noses equipped with sno2 gas sensors. Sensors and Actuators B: Chemical 69(3), 283–286 (2000)
80. Jayaraman, K., Vedula, V.M., Abraham, J.A.: Native mode functional self-test generation for systems-on-chip. In: Proceedings of International Symposium on Quality Electronic Design, pp. 280–285 (2002)
81. Jayaraman, K., Vedula, V.M., Abraham, J.A.: Native mode functional self-test generation for systems-on-chip. In: Proceedings of International Symposium on Quality Electronic Design, pp. 280–285 (2002)
82. Killpack, K., Kashyap, C., Chiprout, E.: Silicon speedpath measurement and feedback into eda flows. In: 44th ACM/IEEE Design Automation Conference, DAC 2007, pp. 390–395 (2007)
83. Killpack, K., Natarajan, S., Krishnamachary, A., Bastani, P.: Case study on speed failure causes in a microprocessor. IEEE Design Test of Computers 25(3), 224–230 (2008)
84. Kim, K.S., Mitra, S., Ryan, P.G.: Delay defect characteristics and testing strategies. IEEE Design Test of Computers 20(5), 8–16 (2003)
85. Koza, J.: Genetic Programming: On the Programming of Computers by Means of Natural Selection. MIT Press (1992)
86. Koza, J.R., Keane, M.A., Streeter, M.J., Mydlowec, W., Yu, J., Lanza, G.: Genetic Programming IV: Routine Human-Competitive Machine Intelligence. Springer, Heidelberg (2003)
87. Kranitis, N., Paschalis, A., Gizopoulos, D., Xenoulis, G.: Software-based self-testing of embedded processors. IEEE Transactions on Computers 54(4), 461–475 (2005)
88. Kranitis, N., Xenoulis, G., Gizopoulos, D., Paschalis, A., Zorian, Y.: Low-cost software-based self-testing of risc processor cores. In: Design, Automation and Test in Europe Conference and Exhibition, pp. 714–719 (2003)
89. Kretzschmar, C., Galke, C., Vierhaus, H.T.: A hierarchical self test scheme for socs. In: Proceedings of 10th IEEE International On-Line Testing Symposium, IOLTS 2004, pp. 37–42 (2004)
90. Krintz, C., Ye, W., Wolski, R.: Application-level prediction of battery dissipation, low power electronics and design. In: Proceedings of the 2004 International Symposium on Low Power Electronics and Design, ISLPED 2004, pp. 224–229 (2004)

91. Krstic, A., Cheng, K.: Delay fault testing for VLSI circuits. Kluwer Academic Publishers (1998)
92. Kuhn, K.: Building predictive models in r using the caret package. Journal of Statistical Software 28(5), 1–26 (2008)
93. Lai, W.-C., Krstic, A., Cheng, K.-T.: On testing the path delay faults of a microprocessor using its instruction set. In: Proceedings of 18th IEEE VLSI Test Symposium, pp. 15–20 (2000)
94. Lai, W.-C., Krstic, A., Cheng, K.-T.: Test program synthesis for path delay faults in microprocessor cores. In: Proceedings of International Test Conference, pp. 1080–1089 (2000)
95. Lamont, G.B., Van Veldhuizen, D.A.: Evolutionary Algorithms for Solving Multi-Objective Problems. Kluwer Academic Publishers, Norwell (2002)
96. Lee, L., Wang, L.-C., Parvathala, P., Mak, T.M.: On silicon-based speed path identification. In: Proceedings of 23rd IEEE VLSI Test Symposium, pp. 35–41 (May 2005)
97. Lindsay, W., Sanchez, E., Reorda, M.S., Squillero, G.: Automatic test programs generation driven by internal performance counters. In: Proceedings of 5th International Workshop on Microprocessor Test and Verification, pp. 8–13 (2004)
98. Liu, C.-N.J., Chang, C.-Y., Jou, J.-Y., Lai, M.-C., Juan, H.-M.: A novel approach for functional coverage measurement in hdl. In: Proceedings of IEEE International Symposium on Circuits and Systems, ISCAS 2000, Geneva, vol. 4, pp. 217–220 (2000)
99. Llobet, E., Brezmes, J., Ionescu, R., Vilanova, X., Al-Khalifa, S., Gardner, J.W., Bârsan, N., Correig, X.: Wavelet transform and fuzzy artmap-based pattern recognition for fast gas identification using a micro-hotplate gas sensor. Sensors and Actuators B: Chemical 83(1-3), 238–244 (2002)
100. Mak, T.M., Krstic, A., Cheng, K.-T., Wang, L.-C.: New challenges in delay testing of nanometer, multigigahertz designs. In: IEEE Design Test of Computers, vol. 21(3), pp. 241–248 (2004)
101. Manetta, L., Ollino, L., Schillaci, M.: Use of an Evolutionary Tool for Antenna Array Synthesis. In: Rothlauf, F., Branke, J., Cagnoni, S., Corne, D.W., Drechsler, R., Jin, Y., Machado, P., Marchiori, E., Romero, J., Smith, G.D., Squillero, G. (eds.) EvoWorkshops 2005. LNCS, vol. 3449, pp. 245–253. Springer, Heidelberg (2005), doi:10.1007/978-3-540-32003-6_25
102. Marcano, D., Duran, F.: Synthesis of antenna arrays using genetic algorithms. IEEE Antennas and Propagation Magazine 42(3), 12–20 (2000)
103. Marco, S., Ortega, A., Pardo, A., Samitier, J.: Gas identification with tin oxide sensor array and self-organizing maps: adaptive correction of sensor drifts. IEEE Transactions on Instrumentation and Measurement 47(1), 316–321 (1998)
104. May, G.S., Spanos, C.J.: Fundamentals of Semiconductor Manufacturing and Process Control. John Wiley & Sons, Inc. (2006)
105. Mayr, E.W.: Toward a new Philosophy of Biological Thought: Diversity, Evolution and Inheritance. Belknap, Harvard (1982)
106. McLaughlin, R., Venkataraman, S., Lim, C.: Automated debug of speed path failures using functional tests. In: 27th IEEE VLSI Test Symposium, VTS 2009, pp. 91–96 (May 2009)
107. Michael, C.C., McGraw, G., Schatz, M.A.: Generating software test data by evolution. IEEE Transactions on Software Engineering 27(12), 1085–1110 (2001)
108. Michael, M.K., Tragoudas, S.: Function-based compact test pattern generation for path delay faults. IEEE Transactions on Very Large Scale Integration (VLSI) Systems 13(8), 996–1001 (2005)
109. Motorola, `http://www.motorola.com`

110. Di Natale, C., Martinelli, E., D'Amico, A.: Counteraction of environmental disturbances of electronic nose data by independent component analysis. Sensors and Actuators B: Chemical 82(2-3), 158–165 (2002)
111. Friedberg, R.M., Dunham, B., North, J.H.: A learning machine: Part ii. IBM Journal 3(7), 282–287 (1959)
112. OpenCores, `http://www.opencores.org`
113. Owens, W.B., Wong, A.P.S.: An improved calibration method for the drift of the conductivity sensor on autonomous ctd profiling floats by theta–s climatology. Deep-Sea Research Part I-Oceanographic Research Papers 56(3), 450–457 (2009)
114. Pacheco, C., Lahiri, S.K., Ernst, M.D., Ball, T.: Feedback-directed random test generation. In: 29th International Conference on Software Engineering, ICSE 2007, pp. 75–84 (May 2007)
115. Padilla, M., Perera, A., Montoliu, I., Chaudry, A., Persaud, K., Marco, S.: Drift compensation of gas sensor array data by orthogonal signal correction. Chemometrics and Intelligent Laboratory Systems 100(1), 28–35 (2010)
116. Padmanaban, S., Tragoudas, S.: Efficient identification of (critical) testable path delay faults using decision diagrams. IEEE Transactions on Computer-Aided Design of Integrated Circuits and Systems 24(1), 77–87 (2005)
117. Paravati, G., Sanna, A., Pralio, B., Lamberti, F.: A genetic algorithm for target tracking in flir video sequences using intensity variation function. IEEE Transactions on Instrumentation and Measurement 58(10), 3457–3467 (2009)
118. Pardo, M., Sberveglieri, G.: Electronic olfactory systems based on metal oxide semiconductor sensor arrays. MRS Bulletin 29(10), 703–708 (2004)
119. Parvathala, P., Maneparambil, K., Lindsay, W.: Frits - a microprocessor functional bist method. In: Proceedings of International Test Conference, pp. 590–598 (2002)
120. Paschalis, A., Gizopoulos, D.: Effective software-based self-test strategies for on-line periodic testing of embedded processors. IEEE Transactions on Computer-Aided Design of Integrated Circuits and Systems 24(1), 88–99 (2005)
121. Pearce, T.C., Shiffman, S.S., Nagle, H.T., Gardner, J.W.: Handbook of machine olfaction. Wiley-VHC, Weinheim (2003)
122. Piziali, A.: Functional Verification Coverage Measurement and Analysis, 1st edn. Springer Publishing Company, Heidelberg (2007)
123. Polster, A., Fabian, M., Villinger, H.: Effective resolution and drift of paroscientific pressure sensors derived from long-term seafloor measurements. Geochem. Geophys. Geosyst. 10 (2009)
124. Pradhan, D.K., Harris, I.G.: Practical Design Verification. Cambridge University Press (2009)
125. Psarakis, M., Gizopoulos, D., Sanchez, E., Reorda, M.S.: Microprocessor software-based self-testing. IEEE Design Test of Computers 27(3), 4–19 (2010)
126. Psarakis, M., Gizopoulos, D., Hatzimihail, M., Paschalis, A., Raghunathan, A., Ravi, S.: Systematic software-based self-test for pipelined processors. In: Proceedings of the 43rd annual Design Automation Conference, DAC 2006, pp. 393–398. ACM, New York (2006)
127. Ravotto, D., Sanchez, E., Schillaci, M., Squillero, G.: An evolutionary methodology for test generation for peripheral cores via dynamic fsm extraction. In: Giacobini, M., Brabazon, A., Cagnoni, S., Di Caro, G.A., Drechsler, R., Ekárt, A., Esparcia-Alcázar, A.I., Farooq, M., Fink, A., McCormack, J., O'Neill, M., Romero, J., Rothlauf, F., Squillero, G., Uyar, A.Ş., Yang, S. (eds.) EvoWorkshops 2008. LNCS, vol. 4974, pp. 214–223. Springer, Heidelberg (2008)

128. Rechenberg, I.: Evolutionsstrategie - Optimierung technischer Systeme nach Prinzipien der biologischen Evolution (PhD thesis) (1971) (Reprinted by) Fromman-Holzboog
129. Robion, A., Sadarnac, D., Lanzetta, F., Marquet, D., Rivera, T.: Breakthrough in energy generation for mobile or portable devices. In: 29th International Telecommunications Energy Conference, INTELEC 2007, pp. 460–466 (2007)
130. Rubinstein, R.Y. (ed.): Simulation and the Monte Carlo method. John Wiley and Sons (1981)
131. Saab, D.G., Saab, Y.G., Abraham, J.A.: Automatic test vector cultivation for sequential vlsi circuits using genetic algorithms. IEEE Transactions on Computer-Aided Design of Integrated Circuits and Systems 15(10), 1278–1285 (1996)
132. Sanchez, E., Sonza Reorda, M., Squillero, G., Violante, M.: Automatic generation of test sets for sbst of microprocessor ip cores. In: 18th Symposium on Integrated Circuits and Systems Design, pp. 74–79 (2005)
133. Sanchez, E., Reorda, M.S., Squillero, G.: On the transformation of manufacturing test sets into on-line test sets for microprocessors. In: 20th IEEE International Symposium on Defect and Fault Tolerance in VLSI Systems, DFT 2005, pp. 494–502 (October 2005)
134. Sanchez, E., Schillaci, M., Squillero, G.: Evolutionary Optimization: the μGP toolkit, 1st edn. Springer, Heidelberg (to be published in July 2011)
135. Sanchez, E., Schillaci, M., Squillero, G., Sonza Reorda, M.: An enhanced technique for the automatic generation of effective diagnosis-oriented test programs for processor. In: Design, Automation Test in Europe Conference Exhibition, DATE 2007, pp. 1–6 (2007)
136. Sanchez, E., Squillero, G., Tonda, A.: Evolution of Test Programs Exploiting a FSM Processor Model. In: Di Chio, C., Brabazon, A., Di Caro, G.A., Drechsler, R., Farooq, M., Grahl, J., Greenfield, G., Prins, C., Romero, J., Squillero, G., Tarantino, E., Tettamanzi, A.G.B., Urquhart, N., Uyar, A.Ş. (eds.) EvoApplications 2011, Part II. LNCS, vol. 6625, pp. 162–171. Springer, Heidelberg (2011)
137. Sanchez, E., Squillero, G., Tonda, A.: Post-silicon speed-path failing-test generation through evolutionary computation. Accepted for Publication in 16th IEEE European Test Symposium, ETS (2011)
138. Schaffer, J.D.: Multiple objective optimization with vector evaluated genetic algorithms. In: Proceedings of the 1st International Conference on Genetic Algorithms, pp. 93–100. L. Erlbaum Associates Inc., Hillsdale (1985)
139. Schwefel, H.-P.: Cybernetic Evolution as Strategy for Experimental Research in Fluid Mechanics (Diploma Thesis in German). Hermann Föttinger-Institute for Fluid Mechanics, Technical University of Berlin (1965)
140. Sharma, R.K., Chan, P.C.H., Tang, Z., Yan, G., Hsing, I.-M., Sin, J.K.O.: Investigation of stability and reliability of tin oxide thin-film for integrated micro-machined gas sensor devices. Sensors and Actuators B: Chemical 81(1), 9–16 (2001)
141. Singh, V., Inoue, M., Saluja, K.K., Fujiwara, H.: Instruction-based delay fault self-testing of processor cores. In: Proceedings of 17th International Conference on VLSI Design 2004, pp. 933–938 (2004)
142. Sisk, B.C., Lewis, N.S.: Comparison of analytical methods and calibration methods for correction of detector response drift in arrays of carbon black-polymer composite vapor detector. Sensors and Actuators B: Chemical 104(2), 249–268 (2005)
143. Smolin, L.: The Life of the Cosmos. Weidenfeld and Nicolson, London (1997)
144. Sánchez, E., Reorda, M.S., Squillero, G.: Test program generation from high-level microprocessor descriptions. In: Reorda, M.S., Peng, Z., Violante, M. (eds.) System-level Test and Validation of Hardware/Software Systems. Springer Series in Advanced Microelectronics, pp. 83–106. Springer, London (2005), doi:10.1007/1-84628-145-8_6

145. Sánchez, E., Reorda, M., Squillero, G.: Efficient techniques for automatic verification-oriented test set optimization. International Journal of Parallel Programming 34, 93–109 (2006); doi:10.1007/s10766-005-0005-7
146. Source Forge. Host of μgp3, `http://sourceforge.net/projects/ugp3`
147. Speek, H., Kerkhoff, H.G., Sachdev, M., Shashaani, M.: Bridging the testing speed gap: design for delay testability. In: Proceedings of IEEE European Test Workshop, pp. 3–8 (2000)
148. Corno, F., Sonza Reorda, M., Squillero, G.: Exploiting the selfish gene algorithm for evolving cellular automata. In: Proceedings of the IEEE-INNS-ENNS International Joint Conference on Neural Networks, IJCNN 2000, vol. 6, pp. 577–581 (2000)
149. Squillero, G.: Microgp - an evolutionary assembly program generator. Genetic Programming and Evolvable Machines 6, 247–263 (2005); doi:10.1007/s10710-005-2985-x
150. Tafertshofer, P., Ganz, A., Antreich, K.J.: Igraine-an implication graph-based engine for fast implication, justification, and propagation. IEEE Transactions on Computer-Aided Design of Integrated Circuits and Systems 19(8), 907–927 (2000)
151. Thatte, S.M., Abraham, J.A.: Test generation for microprocessors. IEEE Transactions on Computers 29(6), 429–441 (1980)
152. Thompson, K.M.: Intel and the myths of test. IEEE Design Test of Computers 13(1), 79–81 (1996)
153. Tomic, O., Eklöv, T., Kvaal, K., Haugen, J.-E.: Recalibration of a gas-sensor array system related to sensor replacement. Analytica Chimica Acta 512(2), 199–206 (2004)
154. Turing, A.M.: Computing machinery and intelligence. Mind 9, 433–460 (1950)
155. van de Goor, A.J.: Testing semiconductor memories: theory and practice. John Wiley & Sons, Inc., New York (1991)
156. Vezzoli, M., Ponzoni, A., Pardo, M., Falasconi, M., Faglia, G., Sberveglieri, G.: Exploratory data analysis for industrial safety application. Sensors and Actuators B: Chemical 131(1), 100–109 (2008); Special Issue: Selected Papers from the 12th International Symposium on Olfaction and Electronic Noses - ISOEN 2007, International Symposium on Olfaction and Electronic Noses
157. Vlachos, D.S., Fragoulis, D.K., Avaritsiotis, J.N.: An adaptive neural network topology for degradation compensation of thin film tin oxide gas sensors. Sensors and Actuators B: Chemical 45(3), 223–228 (1997)
158. Wang, S., Gupta, S.K.: Atpg for heat dissipation minimization during scan testing. In: Design Automation Conference, p. 614 (1997)
159. Weiling, F.: Historical study: Johann gregor mendel 1822-1884. American Journal of Medical Genetics 40(26), 1–25 (1991)
160. Weismann, A.: Evolution Theory. Arnold, London (1904)
161. Wood, B., Milanesi, C., Liang, A., De La Vergne, H.J., Nguyen, T.H., Mitsuyama, N.: Forecast: Mobile terminals, worldwide, 2000-2009. Mobile Communications Worldwide (2005)
162. Yang, K., Cheng, K.-T., Wang, L.-C.: Trangen: a sat-based atpg for path-oriented transition faults. In: Proceedings of the ASP-DAC 2004, Asia and South Pacific Design Automation Conference, pp. 92–97 (2004)
163. Yu, T., Davis, L., Baydar, C.M., Roy, R.: Evolutionary Computation in Practice. SCI, vol. 88. Springer, Heidelberg (2008)
164. Zeng, J., Abadir, M., Bhadra, J., Abraham, J.: Full chip false timing-path identification. In: IEEE Workshop on Signal Processing Systems, SiPS 2000, pp. 703–711 (2000)

165. Zeng, J., Wang, J., Chen, C.-Y., Mateja, M., Wang, L.-C.: On evaluating speed path detection of structural tests. In: 2010 11th International Symposium on Quality Electronic Design (ISQED), pp. 570–576 (2010)
166. Zuppa, M., Distante, C., Siciliano, P., Persaud, K.C.: Drift counteraction with multiple self-organising maps for an electronic nose. Sensors and Actuators B: Chemical 98(2-3), 305–317 (2004)
167. Zurek, W.H.: Decoherence, einselection, and the quantum origins of the classical. Reviews of Modern Physics 75 (2003)

GPSR Compliance

The European Union's (EU) General Product Safety Regulation (GPSR) is a set of rules that requires consumer products to be safe and our obligations to ensure this.

If you have any concerns about our products, you can contact us on ProductSafety@springernature.com

In case Publisher is established outside the EU, the EU authorized representative is:

Springer Nature Customer Service Center GmbH
Europaplatz 3
69115 Heidelberg, Germany

Batch number: 10614940

Printed by Printforce, the Netherlands